THREE MOSQUITOES

Three Mosquitoes

The Biology of Deadly Insects

Jeffrey R. Powell

Johns Hopkins University Press
Baltimore

Printed in the United States of America on acid-free paper
9 8 7 6 5 4 3 2 1

Johns Hopkins University Press
2715 North Charles Street
Baltimore, Maryland 21218
www.press.jhu.edu

Library of Congress Cataloging-in-Publication Data is available.
A catalog record for this book is available from the British Library.

ISBN 978-1-4214-5169-5 (paperback)
ISBN 978-1-4214-5170-1 (ebook)

Special discounts are available for bulk purchases of this book. For more information, please contact Special Sales at specialsales@jh.edu.

EU GPSR Authorized Representative
LOGOS EUROPE, 9 rue Nicolas Poussin, 17000,
La Rochelle, France
E-mail: Contact@logoseurope.eu

To the memories of
George B. Craig Jr.
and
Mario Coluzzi

CONTENTS

PREFACE

> I never write a book unless I can't help it.
> Something has to bother me, like a mosquito, until
> I have to do something to relieve the itch.
>
> —Gregory Maguire

What was it that bothered me, that itch? After running a mosquito research lab for many years, I became aware (and annoyed) that there is no single, modest-sized book that provides an introduction to mosquito research, a book one could recommend to newcomers to the field. But to write a book about mosquitoes in general would be overwhelming and requires expertise far beyond my own—as well as ruling out the "modest-sized" goal.[1] Rather, one of the more remarkable things I've learned over the years is that only three mosquitoes have received the bulk of research studies, one each in three major genera: *Aedes aegypti*, *Anopheles gambiae*, and *Culex pipiens*. While I just used Latin binomials to refer to these three mosquitoes, each is, in fact, a complex of genetically allied but distinct entities of various taxonomic ranks. Work on these three complexes encompasses much of what we know about mosquitoes, and, not coincidently, members of these groups have had the greatest impacts on humans. They have attracted the attention of biologists *because* they are major threats to human health.

Our knowledge of these three mosquitoes is uneven. This is due, at least in part, to how easy or difficult it is to study biological aspects of each species. This inevitably leads to an unequal treatment in the different chapters in this book. I hope that the holes in knowledge apparent from this approach will stimulate efforts to plug them. I must also confess that while I have worked on each of these mosquitoes, my hands-on experience with them is very uneven. I have tried not to let this bias my presentation, but I have no illusions that this has been entirely successful. Also, any expertise I have is in genetics and evolution, so these disciplines form predominant themes. My apologies to physiologists, neurobiologists, and developmental and molecular biologists, all of whom have made fundamental contributions to our understanding of mosquitoes. Other topics I only briefly introduce are the diseases transmitted by mosquitoes and efforts to control them.

Because I hope this book will be useful to a broad range of readers, I have tried to avoid unnecessary jargon and used notes to explain concepts not readily understood by non-professionals. In a sense, I have written this with one thought in mind: *What do I wish I could have learned from a single, modest-sized book when I embarked on my career*? I also use notes to break up the inevitable "dryness" inherent in science writing by relating some interesting fact or anecdote.

It is important to state at the outset that *Three Mosquitoes* is an introduction in the truest sense. I do not provide a thorough review of all relevant literature (which may upset some of my colleagues). Given the ease of computer searches, this is unnecessary, and those interested in more details about any topic can search online to retrieve the most relevant publications, including those appearing after this book was sent to press. I provide the vocabulary needed to efficiently perform such searches.

While most people fear or strongly dislike mosquitoes, I have an inordinate fondness for them,[2] which I have come to find is shared by others. I have met and gotten to know an amazing array of mosquito biologists, among the most diverse set of characters one could hope to have enrich one's life. Fortunately, some of them generously gave of their time and expertise to provide information and figures, share unpublished results, and/or critique parts of this book. I am grateful to Peter Armbruster, Philip Armstrong, Diego Ayala, Dario Balcazar, Nora Besansky, Luciano Cosme, Jacob Crawford, Alessandra della Torre, George Dimopoulos, Dina Fonseca, Michael Fontaine, Didier Fontenille, Andrea Gloria-Soria, Carlos Guerra, Yuki Haba, Ralph Harbach, Julian Hillyer, T. J. Johnson, Stephen Juliano, A. Marm Kilpatrick, Gilbert Le Goff, Tovi Lehamnn, Yvonne Linton, Todd Livdahl, Phil Lounibos, Carolyn McBride, Alistair Miles, Gen Morinaga, Leonard Munstermann, Vincenzo Petrarca, Alex Potter, Nil Rahola, Noah Rose,

Nora Saarman, Igor Sharakhov, Maria Sharakhova, Marta Shocket, Michel Slotman, John Soghigian, Carla Sousa, Andre Wilke, and Andrey Yurchenko. I was most fortunate to have the aid of a fine illustrator, Jacquelyn LaReau, who helped with the 51 figures, the best of which are her original drawings. Andrea Gloria-Soria was particularly generous in quickly providing information and material many times and covering for me during a brief illness. Production of this book was greatly facilitated by the guidance and encouragement of Johns Hopkins University Press editors Tiffany Gasbarrini and Ezra Rodriguez, as well as the meticulous copyediting by Kathleen Capels and final production by Robert Brown. Gisella Caccone generously aided in correcting the proofs. I also need to acknowledge the National Institute of Allergy and Infectious Diseases—in particular its Vector Biology program, so competently run by Adriana Costero–Saint Denis—for financial support over the years.

While many great scientists have devoted their lives to mosquito research, I have dedicated this book to the memories of two who were the greatest influences on me. George B. Craig Jr. instilled in me a love of mosquitoes and personally exemplified the importance of emotional connections in conducting and sustaining a scientific career. In my decades in academia, I have never met any faculty who took such an interest in mentoring and inspiring undergraduate novices. Later, I found a kindred spirit in Mario Coluzzi, who also believed in the importance of evolutionary thinking in understanding mosquitoes and the diseases they transmit. This led to many long lunches over *spaghetti alle vongole* and memorable weekends at Casa delle Palme in Montecelli, Italy.

Notes

1. Shortly after I started this book in 2021, the monumental two volumes of *Mosquitoes of the World* by R. Wilkerson, Y.-M. Linton, and D. Strickman appeared. Unlike the present modest contribution, those 1,308 large pages address the total diversity of known mosquitoes and are an indispensable reference for any professional working on mosquitoes.

2. When the remarkable polymath J. B. S. Haldane was asked by reporters what a lifetime of studying evolution had revealed to him about the Creator, he is said to have replied that He must have an inordinate fondness for beetles (beetles having more described species than any other insect). I cannot find an actual reference to this story, which may well be apocryphal, but the euphonic phrase "inordinate fondness" is memorable.

CHAPTER 1

Taxonomy and History

Ex Africa semper aliquid novi (There is always something new coming out of Africa)

—Pliny the Elder, 23–79 CE

There are more than 3,700 formally named species of mosquitoes (family Culicidae) that occur on every continent except Antarctica. Humans are intimately cognizant of mosquitoes, as we often serve as a blood meal source for these insects. Furthermore, by feeding on our blood, mosquitoes transmit infectious diseases—such as malaria, yellow fever, dengue, filariasis, and many more—that have been the scourge of humans throughout history. Indeed, some of these diseases have played crucial roles in affecting important historical events.[1] While probably fewer than 200 species are of medical importance, mosquitoes have been the subject of intense research for more than 120 years, spurred on by the demonstration (around 1900) that they transmit serious human diseases.

Despite this diversity of mosquitoes, three have received by far the most attention by researchers, for both medical and practical reasons: *Aedes aegypti*, *Anopheles gambiae*, and *Culex pipiens*. While all three were initially given single Linnaean binomial names, as so often happens,

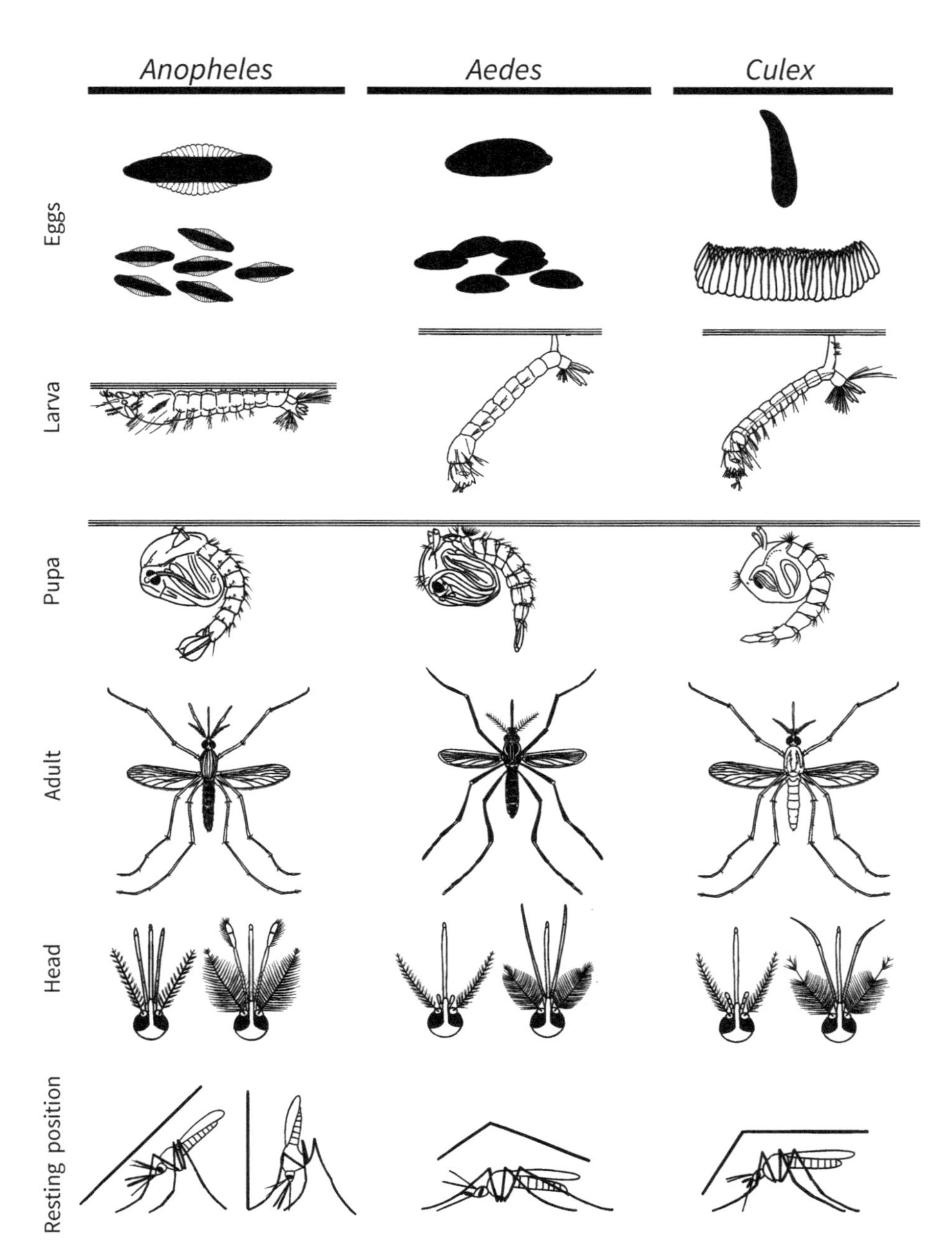

Figure 1.1 Life stages of mosquitoes. **Eggs**: *Anopheles* and *Aedes* lay eggs singly, with the former having "floaters" attached, as these eggs are laid directly on water. *Culex* generally lay eggs in floating rafts of 100–300 eggs. **Larvae**: The most distinctive feature is that *Anopheles* feed parallel to the surface, while *Aedes* and *Culex* hang down, with their breathing tube at the surface. **Head**: In all groups, male antennae are bushier than those of females. **Blood feeding**: Note the distinctive upright posture of *Anopheles*, compared with the other two genera. Details of life stages are in chapters 3–5. Drawing by Jacquelyn LaReau.

intense scrutiny has revealed complexes of related populations that vary in their degree of distinctness, sometimes to the point of requiring recognition as multiple species or other taxonomic units. So, while the title of this book is simply *Three Mosquitoes*, each of these is a heterogeneous amalgam of genetically related groups.

Figure 1.1 is a schematic of the life cycles of typical members of these three genera. Their distinctiveness varies, depending on their life stage. Chapters 3–5 take a closer look at the biology of each of these life stages.

Unraveling Names

Because various names for subdivisions of the three mosquitoes are used in the literature, it is important at the outset to describe how these names are (and have been) used. Issues surrounding the definition of species and other taxonomic units are complex and controversial, and they are not dealt with in this book. Here, the purpose of names is to accurately communicate what the unit being referred to is, how these names have been used in older literature, and how some have changed over time as new information accumulates.

Classical taxonomy relied almost exclusively on morphology to distinguish species and assign latinized binomial names. Between about 1935 and 1950, the biological species concept (BSC) began to supplant the notion that species needed to be morphologically distinguishable. The BSC is an explicitly genetic definition. Species do not exchange genes and thus represent independent evolutionary units. The BSC has been a dominant species concept used to name species in the three mosquitoes considered here. For the *gambiae* complex, given the morphological uniformity of their members, barriers to gene exchange (sterility, inviability, mating discrimination, ecological separation, etc.) have been the primary criteria in describing new taxa. Some named taxa of the *pipiens* complex have minor morphological differences in male genitalia. In contrast, *aegypti* is morphologically highly variable and would probably be considered at least two species under older species concepts, yet these morphologically differentiated types show no barriers to gene exchange.

In this book, I generally drop the genus designations *Aedes*, *Anoph-*

eles, and *Culex* when referring to these three mosquitoes. When *aegypti*, *gambiae*, and *pipiens* are used, this refers to all taxa in the complex—that is, *sensu lato* (s.l.). When referring to the single nominal species of a group, *sensu stricto* (s.s.) is used. As an aid to understanding older and newer literature, below is a brief historical summary of names used for each of the three complexes. This should also help in performing online searches. More details on the genetics of subdivisions within these three mosquitoes are given in chapters 9 and 10.

Aedes aegypti

When most people first hear the name of this mosquito, they immediately ask if it is from Egypt. Linnaeus gave a mosquito from Egypt the name *Culex aegypti* in 1762, but it almost certainly was not the mosquito we refer to today as *Aedes aegypti*. Mattingly (1957) reviewed the history of names applied to this mosquito and rather convincingly concluded that Linnaeus's specimen was most likely *Aedes caspius*.[2]

Ironically, *aegypti* is the morphologically most variable of the three mosquitoes, yet it is the simplest systematically/taxonomically. There are two major forms that are distinct in many ways. They differ morphologically, one being light colored and the other dark. This coloration is due primarily to scales on the abdomen, but the exoskeleton itself also varies in its level of pigmentation. This morphological difference, however, is not a simple dichotomy. Continuous variation abounds and is best documented in the thorough study by McClelland (1974); figure 1.2 is an example. Nevertheless, *in general*, the darker form prevails in sub-Saharan Africa, and there is overwhelming evidence this is the ancestor of the lighter form outside Africa. Because these morphologically distinct types vary geographically, they fit the classic definition of subspecies and thus have been assigned formal names: *Aedes aegypti formosus* is the dark form in Africa, and *Aedes aegypti aegypti* is the lighter form outside Africa. (It is convenient to use Aaf and Aaa to informally designate these two subspecies.) Some workers have used the presence of white scales on the first abdominal tergite as a defining trait for Aaa. This trait follows reasonably well the distinct genetic dif-

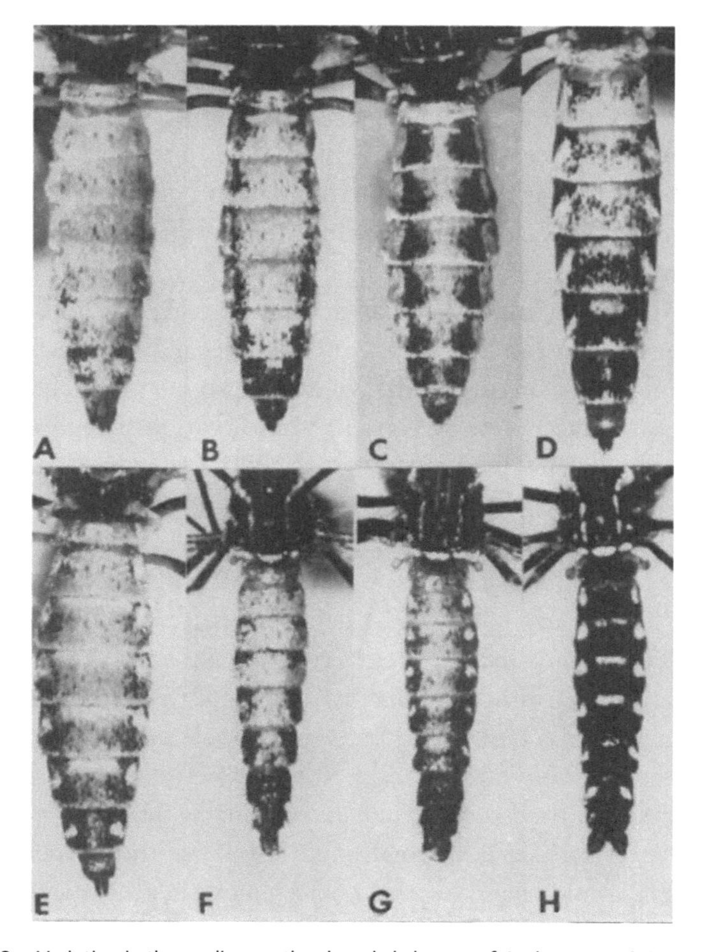

Figure 1.2 Variation in the scaling on the dorsal abdomen of *Aedes aegypti*. **A**, **E**, **F**, and **G** are from Jeddah, Saudi Arabia; **B** from the Philippines; **C** from Ghana; **D** from Malaysia; and **H** from Tanzania. From McClelland (1974).

ferentiation detected by molecular genetic analyses (figure 3-A in Rose et al. 2020).

In addition to morphology, Aaa and Aaf differ in other important ways, such as their larval breeding sites and preferences for blood meals. Generally, Aaf uses natural pools of water (like tree holes) and prefers nonhuman mammals for blood meals. Aaa is adapted to human environments, with larvae found in artificial containers (like discarded tin cans and tires). They generally prefer humans for blood meals. As will become clear, similar to morphology, these are not clear-cut dichotomies.

Importantly, despite their distinctness in a number of ways, there is almost no indication of reproductive isolation between Aaa and Aaf. They readily crossbreed, their offspring are fully fertile, and there is no indication of assortative mating in the lab (D. Moore 1979). Where Aaa and Aaf have reestablished contact in natural conditions—such as on the Kenyan coast in East Africa and in Senegal in West Africa—the mixed populations are in Hardy-Weinberg proportions (Gloria-Soria et al. 2016a). In fact, Jupp et al. (1991) found that offspring from single wild-collected females may vary across the entire morphological range.[3]

A third subspecies of *aegypti*, called *Ae. aegypti queenslandensis*, was described from collections in northern Australia. Mattingly (1957) stated that "it does not differ notably from other parts of the world," except in being "particularly pale." DNA data show it to be identical to Aaa collected in the same locality (Rasic et al. 2016). Thus genetic data do not support this form being a distinct type of *aegypti*. It is best to consider the name *queenslandensis* simply a historical misnomer.[4]

For now, the above is a good guide to the literature and to how names have been used to refer to subdivisions in the *aegypti* complex. A few other points are also worth mentioning, as they may engender confusion. Proposals have been made to rename *Aedes aegypti* as *Stegomyia aegypti* and *Ochlerotatus aegypti*. These names have not been widely adopted, but including *Stegomyia* as a subgenus name is not uncommon, such as in *Aedes* (*Stegomyia*) *aegypti*. More recently, a proposal has been made to do away with all subspecies in the Culicidae and simply raise them to species status (Harbach and Wilkerson 2023). Thus Aaf would be *Aedes formosus*. It remains to be seen how widely this will be accepted.

Anopheles gambiae

While initially described as a single, morphologically uniform species in Africa, over the last 50 years this species has been subdivided into nine formally described species. At first, new species were defined by their incompatibility in crossing: hybrid males are generally sterile, while female hybrids are fertile. For some years, prior to formal latinized species designations, four of the now-recognized species were called A, B, C, and D. The initial six described species are *gambiae* s.s. (species A), *arabiensis* (species B), *quadriannulatus* (species C), *bwambae* (species D), *melas*, and *merus*. Subsequently, Mario Coluzzi and colleagues (Coluzzi et al. 1979, 1985, 2002) documented fixed chromosomal differences among these species, solidifying their status as species under the BSC. In addition to genetic distinctness, there are clear ecological differences. *An. gambiae* s.s. and *arabiensis* are the most widespread, are often closely associated with humans, and account for the majority of malaria transmission in Africa. *An. melas* and *merus* are brackish-water breeders on the west and east coasts of Africa, respectively. *An. quadriannulatus* is widespread and has recently been split into two species: *quadriannulatus* and *amharicus* (Coetzee et al. 2013). These two species take blood meals from nonhuman mammals and are thus not consider to be of much medical importance. Similarly, *bwambae* is a nonhuman animal-biter confined to the Semliki Forest of Uganda. *An. fontenillei*, found in Gabon, has been described as a close relative of *bwambae* (Barron et al. 2019).

An. gambiae s.s. was further subdivided by Coluzzi et al. (1979, 1985), using the terms Mopti, Savanna, Forest, Bamako, and Bissau to designate what they called "chromosomal forms," or sets of populations with distinctive chromosome inversion frequencies, often distinguishable geographically and ecologically. Populations corresponding to two of these chromosomal forms were then found to have a diagnostic characteristic: an easily assayed DNA difference in the intergenic spacer of the rDNA, which Favia et al. (1997) designated as "molecular forms" M (Mopti) and S (Savanna). M was raised to species status (*An. coluzzii*), with S remaining *gambiae* s.s. (Coetzee et al. 2013). Extensive DNA sequencing work has confirmed the genetic distinctness of *coluzzii* and *gambiae* s.s. (see Reidenbach et al. 2012 and chapter 10 for details). The M and S designations are mostly used today as a convenient way to identify field specimens in places where they accurately define *coluzzii* and *gambiae* s.s.

There is recent evidence the chromosomal form Bissau is genetically distinct and may warrant formal taxonomic recognition (Caputo et al. 2024). The taxonomic status—or even validity—of the other chromosomal forms (Forest and Bamako) is unclear at present.

Figure 1.3 is a rough timeline, indicating the changes in nomenclature for the *gambiae* complex. Note that many of these subdivisions have only been proposed quite recently. Genome sequencing has documented further genetic distinctions that may warrant the recognition of more taxa (e.g., Tennessen et al. 2021; Caputo et al. 2024) or, perhaps equally likely, given the rates of hybridization between some named subunits (e.g., Fontaine et al. 2015; Thawornwattana et al. 2018), taxon designations may collapse. At present, it seems safe to assume that the formal latinized names in italics in figure 1.3 are valid and likely to remain in some iterations.

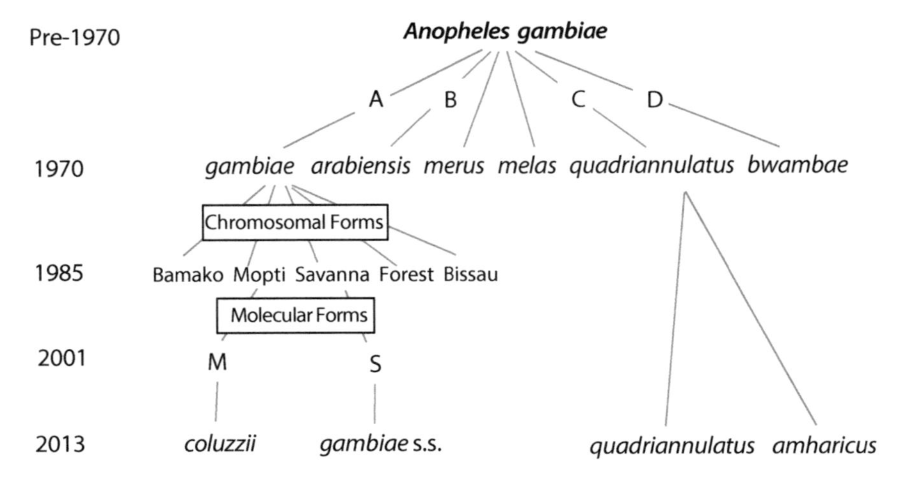

Figure 1.3 Timeline for names applied to subdivisions of the *Anopheles gambiae* complex. Until about 1970, a single name was used, followed by breaking up the taxon into six species, where Latin designations for the two saltwater breeders (*melas* and *merus*) and A, B, C and D for the others were used for some time. In 1970, these lettered designations were assigned latinized species names. A confusing period followed, when *gambiae* s.s. was broken down into "chromosomal forms" and "molecular forms." Today, the eight italicized names are the accepted species designations, although older literature uses other names.

Culex pipiens

The common name for this mosquito is the "house mosquito." It is distributed worldwide in both temperate and tropical climes. Like *aegypti* and *gambiae*, *pipiens* was initially considered to be a single species. Two subspecies were then described—a largely temperate form, *C. pipiens pipiens*, and a tropical form, *C. pipiens quinquefasciatus*. More recently, both were raised to species status as *C. pipiens* s.s. and *C. quinquefasciatus*. These species differ morphologically in minor ways in male genitalia, but, more notably, in an important physiological trait—the ability to survive winters in diapause.[5] *Culex pipiens* s.s., having a predominantly temperate distribution, can diapause, while tropical *quin-*

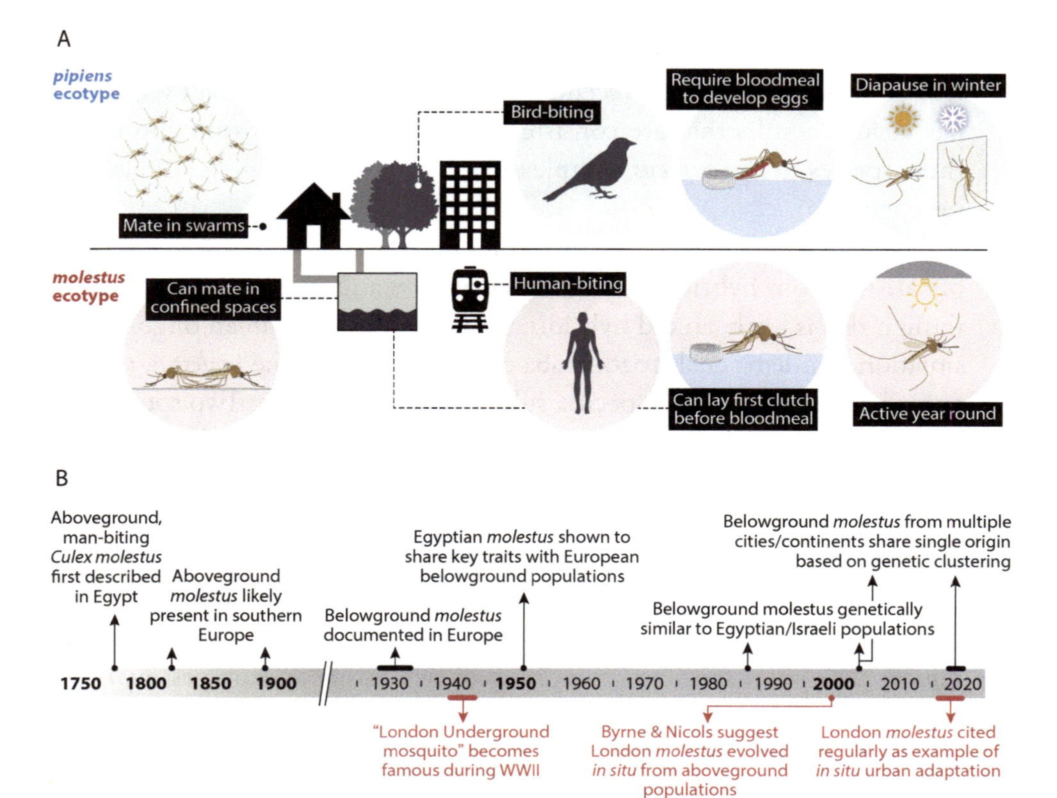

Figure 1.4 **A.** Depiction of differences between ecotypes of *Culex pipiens*. **B.** A proposed timeline for their evolution. From Haba and McBride (2022).

quefasciatus cannot. (The names *C. fatigans* or *C. pipiens fatigans* were also used for *quinquefasciatus* and often appear in the literature before about 1975.)

In some urban settings in both the United States and Europe, *pipiens* has become closely adapted to subterranean human habitats, including subways. These populations differ from "normal" *pipiens* in being autogenous (able to produce eggs without a blood meal) and not diapausing. They have been assigned another Latin name, *molestus*, which is sometimes used for a species, subspecies, ecotype, or biotype.[6] Figure 1.4 summarizes the differences between *pipiens* s.s. and *molestus*.

Both allozyme and microsatellite studies have confirmed the genetic distinctness of *pipiens* s.s. and *quinquefasciatus* (e.g., Fonseca et al. 2004; Weitzel et al. 2009; Becker et al. 2012). This last citation has excellent photographs illustrating the different ecologies and male genetalia of these species. Most recently, whole genome sequencing (WGS) has yielded results that are consistent with the genetic distinction of these species in the *pipiens* complex (e.g., Yurchenko et al. 2020; also see chapter 9).

The name *pallens* has been used for a form of *pipiens* in Asia, possibly arising from hybridization between *pipiens* and *quiquefasciatus*, although this is likely an old hybridization event rather than an ongoing situation (Aardema et al. 2020; Haba et al. submitted). Like *molestus*, the name has been used for a species, subspecies, and biotype. Two species native to Australia are more distinct: *C. australicus* and *C. globocoxitus*. They appear to be paraphyletic clades derived from *quinquefasciatus*.

The foregoing very brief treatment is a guide to how names have been applied to *C. pipiens* s.l. and should help readers understand literature that has used these names in various ways. Today the accepted nomenclature is to designate three species: *quinquefasciatus*, *pallens*, and *pipiens*. The latter species is comprised of two forms or ecotypes: *pipiens* s.s. and *molestus*. Farajollahi et al. (2011) and Aardema et al. (2020) are recent treatments of the complex systematics, taxonomy, and phylogeny of *pipiens* s.l., with the former paper also discussing the role of members of the complex in disease transmission.[7]

Ancient History

Mosquitoes have existed for at least 100 million years (Grimaldi and Engels 2005) and perhaps as long as 220 million years (Reidenbach et al. 2009; Soghigian et al. 2023). The closest extant relatives of mosquitoes are "phantom" midges in the family Chaoboridae. Other close relatives, such as black flies and biting midges, also rely on blood meals from vertebrates. For this reason, taking blood meals is likely an ancestral trait predating the origin of Culicidae. Blood feeding has been lost several times in these lineages, as well as within Culicidae.

While Culicidae has various subfamilies and a somewhat complex taxonomy (Harbach 2007), the majority of common species fall into three genera: *Anopheles*, *Aedes*, and *Culex*. Thus the titular *Three Mosquitoes* conveniently represent much of the diversity of Culicidae. There is little doubt that the phylogenetic relationship of these three genera is as depicted in figure 1.5. Placing absolute time estimates on the nodes in this small tree is more controversial than the structure of the tree

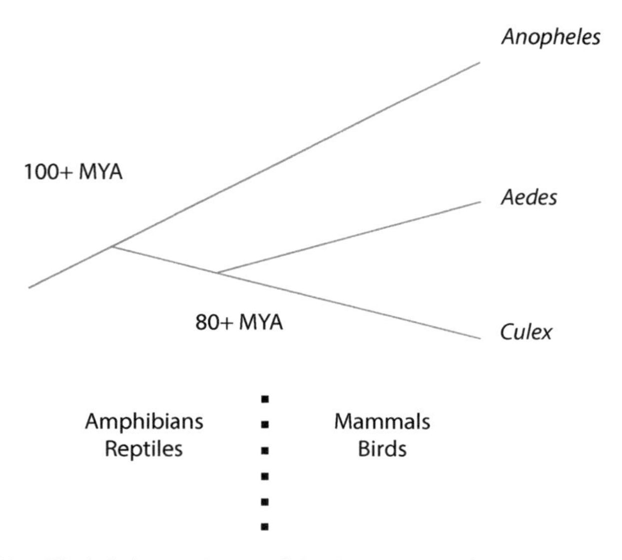

Figure 1.5 Simplified phylogenetic tree of the three genera of mosquitoes discussed in this book. Minimal dates are noted on the nodes, as are the dominant vertebrates present at the time.

itself, as the only mosquito fossils are encased in amber. Poiner et al. (2000) made a critical evaluation of all claims and concluded that 15 fossils could be rigorously assigned to Culicidae, the oldest being 90–100 MYA (million years ago), in the mid-Cretaceous (Borkent and Grimaldi 2004). It is highly unlikely that any usable DNA can be extracted from amber-preserved insects more than a few hundred years old (Lalonde and Marcus 2020; Peris et al. 2020), contrary to claims in the movie *Jurassic Park*.[8]

There is little doubt that anophelines split from the other two genera at least 100 MYA, and possibly as early as 220 MYA. Aedinies split from culicines at least 70 MYA, and possibly up to 200 MYA. The important point is that *mosquitoes are ancient insects* that have been around longer than most extant vertebrates. Blood feeding, being an ancestral trait in mosquitoes, almost certainly started on reptiles and amphibians, as mammals and modern birds (their major blood sources today) would have been scarce or absent when mosquitoes arose. An extensive analysis of blood feeding, based on phylogenetic trees, confirms that amphibians are the most likely ancestral source (Soghigian et al. 2023; also see chapter 6).

Distributions and Recent Histories

The distributions and histories of mosquito species are closely related. The fields of biogeography and, more recently, phylogeography aim to connect these two attributes of species. Various sources of information bear on these subjects, ranging from geology to molecular genetics, as well as to entomological collection records and epidemiology (times and places where mosquito-borne diseases are documented). These sources of information are very uneven for the three mosquitoes under consideration. Perhaps what is most remarkable is that one of these mosquitoes has a rather stable distribution (*gambiae*), while the other two have complex histories and distributions that continue to change.

Aedes aegypti

Figure 1.6 shows the present-day distribution of the two subspecies of *aegypti*, including regions where the two have recontacted on the east

Figure 1.6 Present distributions of the subspecies of *Aedes aegypti*. Black indicates the ancestral subspecies *formosus*, and dark gray is used for the derived *aegypti*. Light gray in West Africa and coastal Kenya denotes the presence of both, often as hybrids. The light gray area in Argentina indicates the only populations outside Africa retaining a partial African (Aaf) genetic signature (see figure 9.1). The light gray in Madagascar indicates ancestral taxa of various taxonomic affinities that are yet to be clearly defined. Drawing courtesy of Luciano Cosme.

and west coasts of Africa. It also notes the unique status of Argentinian populations retaining considerable African ancestry. In addition, Madagascar and surrounding islands contain ancestral populations, as well as closely related species, some of which are undescribed. The historical events that gave rise to this distribution are quite well understood. Figure 1.7 summarizes our present understanding of how and when this distribution was established.

Ae. aegypti belongs to the subgenus *Stegomyia*, which almost certainly had its origin in Asia. Around 60 MYA, a lineage broke off and colonized islands in the southwestern Indian Ocean, Madagascar, and smaller islands off the east coast of Africa. (This occurred after Madagascar split from Asia, about 85 MYA.) Two lineages then entered continental Africa. Around 30 MYA, one lineage colonized continental Africa, giving rise to present-day *Ae. bromelliae*, *Ae. simpsoni*, *Ae. africanus*, and related species. The lineage that gave rise to *Ae. aegypti* remained within the southwestern Indian Ocean area and did not colonize the continent until 50,000–80,000 years ago, during a relatively long, wet

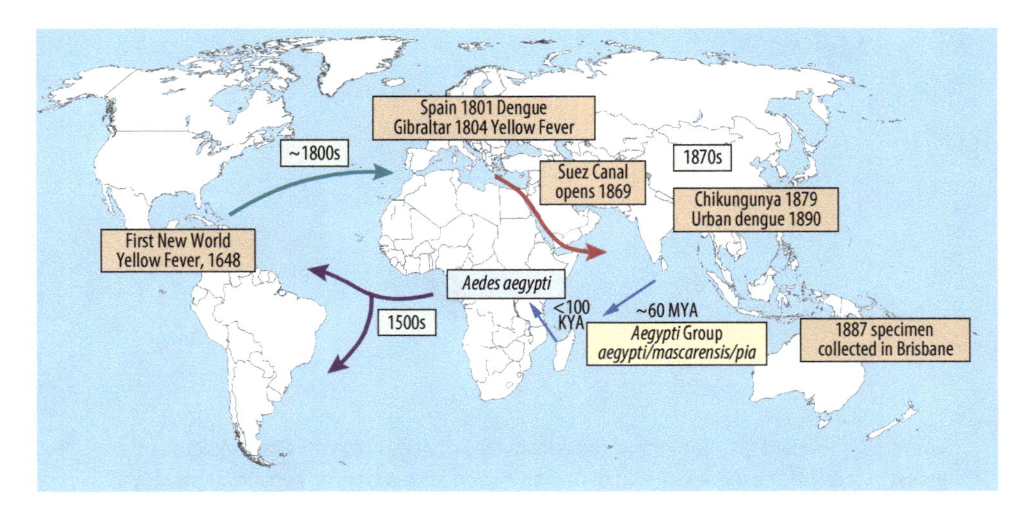

Figure 1.7 Events leading to the establishment of the *Aedes aegypti* distribution shown in figure 1.6, based primarily on genetic data consistent with the epidemiological and historical events noted. From Powell et al. (2018), with additional information on the role of southwestern Indian Ocean islands documented in Soghigian et al. (2020).

period in the history of East Africa (Soghigian et al. 2020). This lineage then spread across sub-Saharan Africa.

The next significant event occurred in West Africa. Prolonged dry periods of almost no rain for up to six months are common—a stressor for a humid, forest-adapted mosquito. Natural larval breeding sites would be dry, and gravid females looking for a place to lay their eggs in order to get through the dry season would have found water stored by villagers to be very attractive places to oviposit. Females eclosing (emerging from pupae as adults) from human-stored water sites would find humans to be the most inviting source for blood meals and thus developed a taste for humans. The likely role of aridity in mosquitoes' evolving preferences for humans is supported by Rose et al. (2020) and discussed in chapter 6. Depending on locality, permanent human villages began appearing in sub-Saharan Africa 4,000–10,000 years ago.

Shortly after the New World was "discovered" by Europeans, an extensive slave trade ensued. Ships would leave their European ports and travel down the west coast of Africa to pick up natives destined for the New World, where they provided crucial labor for burgeoning agricultural enterprises, especially sugar and cotton plantations. (Eltis and Richardson 2010 is a wonderfully illustrated guide to the transatlantic slave trade). Before embarking on the two- to four-month cross-Atlantic trip, ships would also stock up on fresh water, which not infrequently contained *aegypti* larvae. These populations had already adapted to human-generated water containers and evolved a preference for human blood meals, which preadapted them for life aboard ships. Upon arrival in the Americas, they would have found several native species of mosquitoes adapted to the forest niche that *aegypti* occupied in Africa. Instead of competing, these partly domestic populations, designated "proto-Aaa" by Powell et al. (2018), continued to breed in human-generated environments and eventually became the fully domesticated form we recognize as Aaa. Given the patterns of the slave trade, as well as genetic relationships, the region around what is today Angola is the likely source of the introductions. This mosquito also brought yellow fever—the major, easily recognized disease being transmitted by *aegypti* at this time—which was first reported in the New World in the 17th century.

It is less clear how Asia became colonized by *aegypti*. One might assume that the East African populations crossed the Indian Ocean during periods of extensive trade, especially with India. All genetic data, however, indicate that Asian populations are derived from New World populations. This might have occurred via Pacific Ocean trade from the west coast of South America, but it is more likely that, just as the slave trade introduced *aegypti* to the Americas, when these ships returned to their European ports of origin, they brought this mosquito with them. Breeding populations of *aegypti* existed in the Mediterranean basin from about 1800 to 1950.[9] The opening of the Suez Canal in 1869 provided a shipping corridor into Asia, consistent with the first reports of urban dengue in Asia in the 1870s. Given the close relationship of Black Sea *aegypti* to the extinct Mediterranean populations (figure 9.5), however, overland transport to Asia cannot be discounted. Colonization of Australia and the South Pacific Islands followed.[10]

A significant contraction of *aegypti* populations in the New World took place due to an eradication campaign initiated after World War II, when DDT was developed. The goal was to eradicate this mosquito

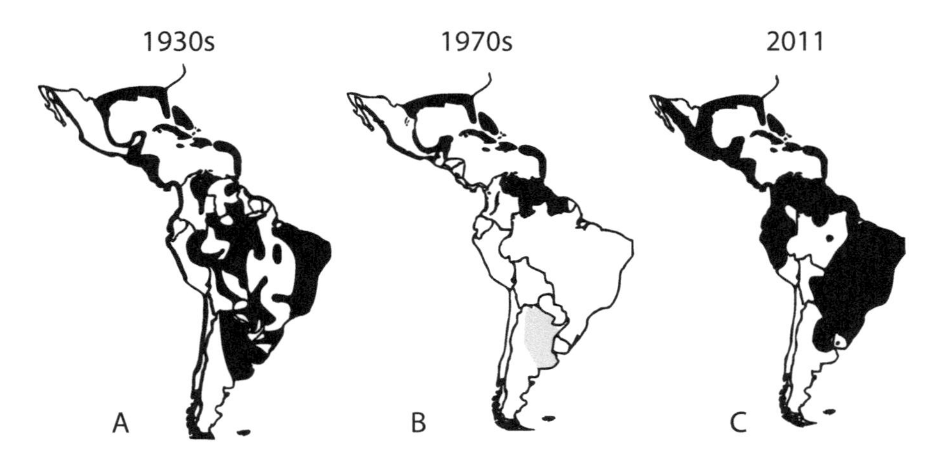

Figure 1.8 Effects in the Americas of a campaign to eradicate Aaa after World War II. **A**, **B**, and **C** are official WHO maps. **B** is modified to reflect the retention of breeding populations in Argentina (in gray) throughout the eradication program, as documented by Savero (1955). Contemporary *Aedes aegypti* populations in the region retain a unique genetic signature (see figure 9.1).

from the invaded New World territory. This was partially successful up to about 1970, when the campaign was halted, due to the adverse environmental damage caused by widespread use of DDT became apparent, as well as widespread use of a yellow fever vaccine, which had reduced health concerns. Shortly after this cessation, *aegypti* regained its former distribution (figure 1.8). Importantly, in South America there is evidence that eradication was not achieved in northwestern Argentina and southern Bolivia and Paraguay. Population-genetics evidence presented in chapter 9 supports the hypothesis that present-day populations in this region may represent a refugia for the original introduction of *aegypti* from Africa.

It is of some interest in considering *aegypti*'s invasive history to ask why, out of the hundreds of species of mosquitoes in sub-Saharan Africa (59 in the subgenus *Stegomyia*) only one became human adapted and hitchhiked with humans around the world? One possibility is that *aegypti*, in its native habitat, tends to breed at the edge of forests, in ecotones with open savanna (Lounibos 1981). This means that as humans built villages on savannas or cut down forests to clear land for a village, *aegypti* was the closest forest mosquito most likely to wander into villages in the dry season and find suitable oviposition sites, thus setting the domestication process in motion.

The above discussion provides an overview of the major events in the history of *Ae. aegypti* and accounts for the distribution of year-round permanent populations (figure 1.4). While their permanent distribution is temperature restricted, during warmer summer months the species can temporarily expand into higher latitudes (Hahn et al. 2016, 2017). For example, Philadelphia experienced a particularly severe yellow fever outbreak in 1793, with about one tenth of the population dying in the then-largest city in the newly formed United States. *Ae. aegypti* continues to move out of its permanent range every summer and even occasionally establish permanent breeding populations (Powell 2016; table 9.3).

Anopheles gambiae s.l.

While all members of the *gambiae* complex are found only in Africa, the various taxa have complex geographical relationships (figure 1.9; Coetzee et al. 2000). Most members of the complex breed in stagnant water, especially when it is sunlit. This is unlike *aegypti*, which prefers shaded standing water. Thus larval competition between *aegypti* and *gambiae*, whose distributions overlap, is minimal. Two species of the *gambiae* complex breed in brackish salt water, one on Africa's east coast (*merus*) and one on the west coast (*melas*), although *merus* occurs in brackish water located considerably inland. Larvae of these species have one of the few distinguishing characters (distinctive larval pectin; see figure 4.1) in an otherwise morphologically uniform group of species. *An. bwambae* has the narrowest distribution, as it is only found in the Semliki Forest of Uganda.

The entire complex likely arose in Africa, as its closest relatives are all in that continent. The least understood aspect of the complex's history and distributions is the timing of crucial events. Coluzzi (1982) argued that much of the diversification resulting in speciation occurred

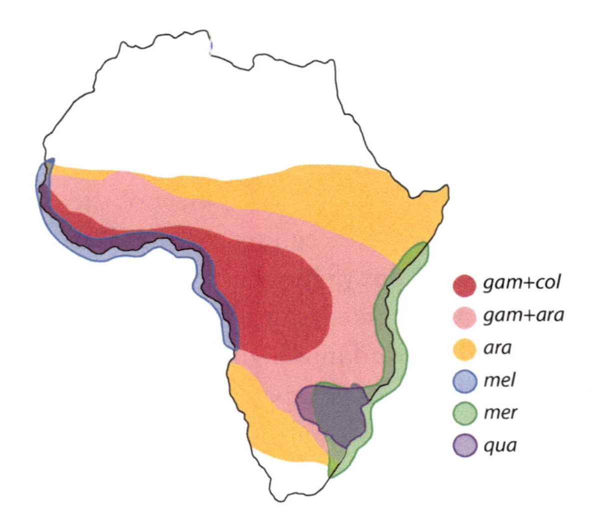

Figure 1.9 Distribution of members of the *Anopheles gambiae* complex. The abbreviation gam is *gambiae* s.s., col is *coluzzii*, ara is *arabiensis*, mel is *melas*, mer is *merus*, and qua is *quadriannulatus*. From Fontaine et al. (2015).

in response to changes induced by humans 4,000–10,000 years ago.[11] Present ecological niches of members of the complex include irrigated rice fields or hoofprints left by domestic animals. Using DNA data to date key events in the history of *gambiae* suggest older time periods for their diversification (chapter 10).

Unlike the other two mosquitoes dealt with here, *gambiae* has remained relatively stable in its distribution, despite its close association with humans. With one notable exception, it has never formed stable populations outside Africa. In 1930 it was detected in northeastern Brazil, around the city of Natal. Thanks to ongoing mosquito surveillance monitoring yellow fever and *aegypti*, this introduction was quickly detected, although it took 10 years to fully eradicate *gambiae* from Brazil (Soper and Wilson 1943).[12] Malaria in the region greatly increased during that time. In the 1930s, *An. gambiae* was still considered a single species. Only recently, using DNA extracted from museum specimens collected during the invasion, was it determined that the Brazil invader was *An. arabiensis*, the most arid-adapted member of the complex (Parmakelis et al. 2008).

Culex pipiens s.l.

The distributions of the various taxa in this complex are shown in figure 1.10. Like the other two mosquitoes, Africa is the likely origin of *pipiens*, based on the presence of closely related species, although Asia is also a possible origin for this complex. Contemporary East African and Asian populations of *quiquefasciatus* have the highest allelic diversity, implying that they are the oldest populations (Fonseca et al. 2006). This probably indicates that a tropical or subtropical form, like *quinquefasciatus*, was ancestral. More-temperate *pipiens* may have originated in northern Africa—perhaps in Ethiopia, as an adaptation to cooler, higher elevations (Barr 1967)—and subsequently colonized Europe. Given that, as far as is known, *quinquefasciatus* and *pipiens* are each monophyletic, their colonization in the New World must have involved separate events by temperate *pipiens* from Europe and tropical *quinquefasciatus* from Africa or Asia. The timing of these events is likely similar to that for

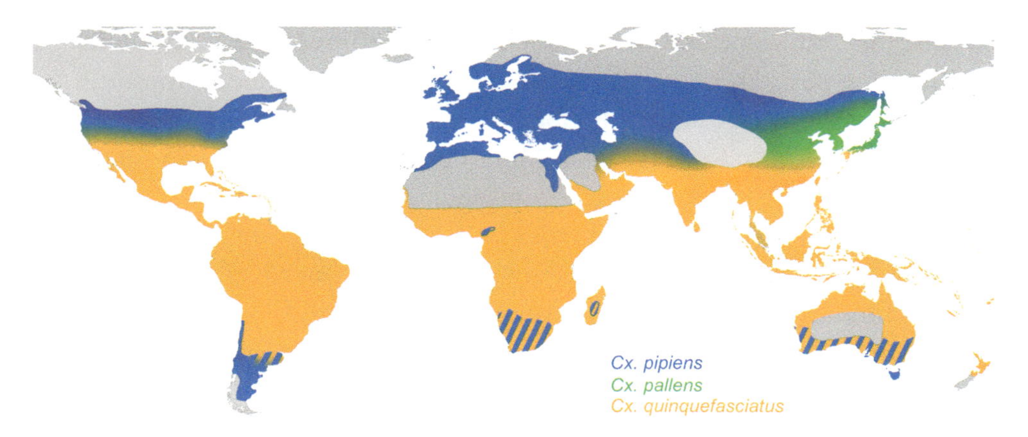

Figure 1.10 Global distribution of three members of the *Culex pipiens* species complex: *pipiens* s.s. (blue), *pallens* (green), and *quinquefasciatus* (yellow). Fuzzy shading where the colors meet indicates areas where hybrids are often found. Two morphologically indistinguishable ecotypes/forms of *pipiens* s.s. co-occur throughout much of the Northern Hemisphere, where *molestus* is often (but not always) segregated in human-made belowground habitats. In other geographic areas, one or the other ecotype occurs in isolation—such as *pipiens* in sub-Saharan Africa and *molestus* in Egypt (where it likely originated), as well as in Australia and parts of Argentina (where it has been introduced in modern times). Figure kindly provided by Carolyn MacBride.

Ae. aegypti, being tied to patterns of human transport and movements. A major difference from *aegypti* is that Asian populations of *pipiens* are older than New World populations, as are Australian *pipiens* populations. New World populations originated relatively recently, consistent with their low genetic diversity.

The distributions of *pipiens* and *quinquefasciatus* closely track temperature, as was demonstrated by a study of a unique situation in California. Due to the physiography of the landscape, temperatures are warm in the far southern part of the Central Valley, become gradually cooler moving north, then become warm again, followed by cooling even farther north. This inversion of the usual north-south temperature gradient over ~200 km is followed by the relative frequencies of *pipiens* and *quinquefasciatus* (Tabachnick and Powell 1983; a follow-up study by Urbanelli et al. 1997 largely confirms these findings).

The origin(s) and historical biogeography of the *molestus* form are more controversial. Aardema et al. (2020) and Yurchenko et al. (2020) interpreted their data as indicating independent origins for *molestus* populations in Europe and the Americas, derived from local aboveground breeding populations of *pipiens* s.s. This suggests that *molestus* mosquitoes in the Underground in London are no older than the Underground itself (dating to 1863). Haba and McBride (2022) reviewed a large literature and provided new data they interpret as indicating that a human-biting form of the *pipiens* complex (= *molestus*?) evolved thousands of years ago in the Middle East and subsequently spread. (*C. pipiens* is predominantly a bird-biting mosquito; see chapter 6.) Part of the confusion likely lies in the fact that the term *molestus* is not clearly defined. For example, as indicated in figure 1.4, three characteristics of *molestus* today are human biting, autogeny, and lack of diapause. It is conceivable that these three traits did not arise simultaneously. Haba and McBride (2022) focused on blood-meal choice, which may have arisen before the stegnomy (ability to mate in confined spaces) and autogeny of today's belowground-breeding *molestus*. Fonseca et al.'s (2004) molecular data produced a phylogenetic tree with *molestus* from Asia, Australia, and Europe forming a monophyletic group, consistent with a single origin (chapter 9). Hybridization between *pipiens* s.s. and *molestus* is discussed in chapter 10.

Asian *C. pallens* is hypothesized to have originated as a hybrid between *pipiens* and *quiquefasciatus*. The Australian endemics *C. globocoxitus* and *C. autralicus* are recent paraphyletic offshoots of Australian *quinquefasciatus*.

Aardema et al. (2020), Haba and McBride (2022), and Haba et al. (submitted) are recent key papers on which much of the foregoing is based.

Notes

1. Several professional historians have considered the role of mosquito-borne diseases in affecting the outcome of important historical events. Among the more intriguing explications is John R. McNeill's *Mosquito Empires* (2010) recounting the major role yellow fever (transmitted by *Aedes aegypti*) played in the European invasion of the New World. Malaria has also played a major role in history, dating back at least to Roman times. Shah's *The Fever: How Malaria Has Ruled Humankind for 500,000 Years* (2010) is a readable introduction to a much larger literature.

2. The number of names applied to what we recognize today as *Ae. aegypti* is so great that, as Christophers (1960) stated in his magisterial book, until 1900, " the trouble was that it had so many aliases, almost one for every country and systematist." He was able to find 24 latinized names that had been used!

3. The one study indicating some reproductive isolation between geographic strains of *aegypti* is Dickson et al. (2016). In laboratory crosses, these authors found reduced oviposition and some degree of lowered viability in hybrids with Senegal strains of Aaf and older lab strains of Aaa. Little follow-up work has been done to confirm this, and it is not clear how these lab-based studies relate to what occurs in the field.

4. With regard to scaling variation, McClelland (1974) concludes (with tongue in cheek?), "The only objection to the use of the var. *queenslandensis* is that it is of little value as a label since it covers about six-sevenths of the range of variation of the species."

5. Diapause is a state of suspended activity in adult mosquitoes, especially females, that is an adaptation allowing them to survive several months of otherwise inhospitable climates, most often cold weather. A number of physiological changes are induced by the shortening of day length and decreasing temperatures. Female *pipiens* s.s. take a final blood meal before entering diapause. This is not used to make eggs, but to produce lipids that support a low metabolism throughout the cold months. An additional blood meal is usually required after diapause to produce eggs. See chapter 7 for more on diapause in *pipiens*.

6. "Biotype" is used to designate a distinct subdivision of a species that has particular attributes. It is sometimes equated with genetic distinctness, but usually some additional biological attribute (like autogeny) is involved. Similarly, "ecotype" usually designates some distinct ecological feature, such as where a group of populations is breeding (e.g., aboveground or subterranean). "Form" is a purposely ambiguous term, simply meaning that there is some distinction but making no reference to what the distinction is. It does not have any taxonomic implications.

7. A species name, *Culex juppi*, has been given to populations in South Africa that are distinguished by a lack of *Wolbachia* infections (chapter 11) and distinct mtDNA haplotypes (Dumas et al. 2016). The taxonomic status of this species is unclear, as there is no formal description or type specimen.

8. Another problem with the premise of this movie is that the amber-enclosed mosquito pictured during a tour of the visitors' center is almost certainly a member of the genus *Toxorhynchites*, which does not take blood meals.

9. In a letter written in 1801, the queen of Spain described suffering an illness that was very likely *aegypti*-transmitted dengue fever, although she had not traveled outside Spain (Rigau-Pérez 1998). She used the Spanish term *quebranta huesos* (breakbone fever), which became a common name for dengue. Gibraltar suffered severe dengue outbreaks in 1800–1804, and dengue in Greece was particularly deadly in 1927–1928 (Rosen 1986; Sawchuyk and Burke 1998). The use of DDT after World War II to control malaria around the Mediterranean, coupled with increasing use of indoor plumbing, caused Mediterranean *aegypti* to become extinct by about 1950. Relics of these populations exist around the Black Sea today (Kotsakiozi et al. 2018a; figure 1.6).

10. Why are Asian *aegypti* not derived from populations on the east coast of Africa, a much more direct route than the circuitous trip to the New World and back through the Mediterranean? This is particularly puzzling, as there was extensive Asian trade with the east coast of Africa, especially from India, dating back to at least 200 CE. The answer may be that dry seasons in East Africa are not as prolonged as those in West Africa. Thus *aegypti* was not pushed into becoming domesticated and instead remained a sylvan mosquito, unlikely to become a stowaway on trading vessels. One piece of evidence in favor of this is that, while yellow fever existed in forested localities in East Africa for several centuries (at least up until 2008), there were no reported cases of yellow fever being transmitted in towns or cities (Ellis and Barrett 2008). Genetic studies indicate that, beginning in the mid-20th century, domestic-breeding Aaa in East Africa likely are derived from already established Asian populations.

11. A summary of these arguments, and data supporting them, include Coluzzi (1982), Coluzzi et al. (1979, 1985), and Touré et al. (1998a, 1998b). Mario Coluzzi was a remarkable researcher who did more to unravel the many-faceted nature of the *An. gambiae* complex in Africa than anyone else. His energy, devotion, and generosity inspired a large number of researchers, especially native Africans. Powell et al. (2014) is a memorial biography of Coluzzi.

12. The potential disaster that was avoided was dramatized by the Rockefeller Foundation's 1938 annual report: "*Anopheles gambiae* is potentially a much more dangerous

invader than the Martians would have been. H. G. Wells's Martians, it will be remembered, were unable to adjust themselves to life on this planet and quickly died. *Anopheles gambiae* striking from equatorial Africa, has invaded South America and is making itself very much at home in Brazil."

CHAPTER 2

Genomes

> The genome is a book that wrote itself, continually adding, deleting and amending over four billion years.
>
> —Matt Ridley

In many ways, the three mosquitoes featured in this book are quite similar, such as in their morphology and life stages (see figure 1.1). One might expect this similarity be reflected in their genomes. But, as will become clear in this chapter, the genomes of these three mosquitoes are very different from one another in both structure (size and composition) and function (e.g., sex determination). Each genome is unique, with its own idiosyncrasies.

Genome Sizes

Table 2.1 summarizes various properties of the genomes of the three mosquitoes. Their overall haploid genome size varies from 260 Mb (megabases, or millions of DNA base pairs) for *gambiae* to 1,220 Mb for *aegypti*.[1]

What causes this variation in genome size? Examination of table 2.1 and figure 2.1A quickly reveals that the variation among these species is due to repetitive DNA sequences. Repetitive sequences differ considerably in both their structure and how they may have arisen. What is called

Table 2.1 Summary of genomes of three mosquitoes

Group	Physical size	Protein-coding genes	Nonprotein-coding genes	Repetitive	Transposable elements	Total genomic recombination	cM/Mb DNA
aegypti	1,220 Mb	14,718[1]	5,086	65%[2]	55%	205–230 cM[3]	0.17
gambiae	260 Mb[4]	13,094	738	10%–20%	11%–16%	222 cM[5]	1.08
pipiens	560 Mb[6]	15,081[6]	828	61%	29%	186 cM[7]	0.72

Note: Data come from VectorBase.org, except where noted. Nonprotein-coding genes include tRNAs, rRNAs, and the like. Abbreviations: cM = centimorgans; Mb = megabases.

1. B. Matthews et al. (2018).
2. 55% transposable elements, 3.3% simple tandem repeats, 7% unclassified repeats.
3. Severson et al. (2002); Juneja et al. (2014).
4. Besansky and Powell (1992). Zamyatin et al. (2021) report genomes for multiple species of this group, ranging up to 273 Mb for *coluzzii*.
5. Dimopoulos et al. (1996); L. Zhang et al. (1996).
6. Liu et al. (2023); Ryazansky et al. (2024).
7. Hickner et al. (2013).

"satellite DNA" consists of tandem arrays of thousands of the same 100–400 bp (base pair) sequences. Generally, these sets of repetitions are concentrated near the centromere (a constricted region of a chromosome where spindle fibers are attached during cell division), as in these mosquitoes, and are called "heterochromatin."[2] Longer, interspersed repetitive sequences exist throughout the genomes. These are variously called SINES and LINES (short and long interspersed sequences) and are thought to have their origin as genomes of retroviruses that, when they reverse transcribe RNA into DNA, became inserted in their host's genome. Some of these sequences can continue to move in a genome and are called "transposable elements," or TEs. The *aegypti* genome is particularly interesting in this regard. Fully 58% of the genome consists of TEs, although only a minority of them are active.

Protein-Coding Genes

While the genome size varies considerably among the species, it is remarkable that the number of protein-coding genes (most of the single-copy fraction of genomes) is quite uniform across species, ranging from 13,000 to 15,000.[3] These protein-coding genes have various lengths, al-

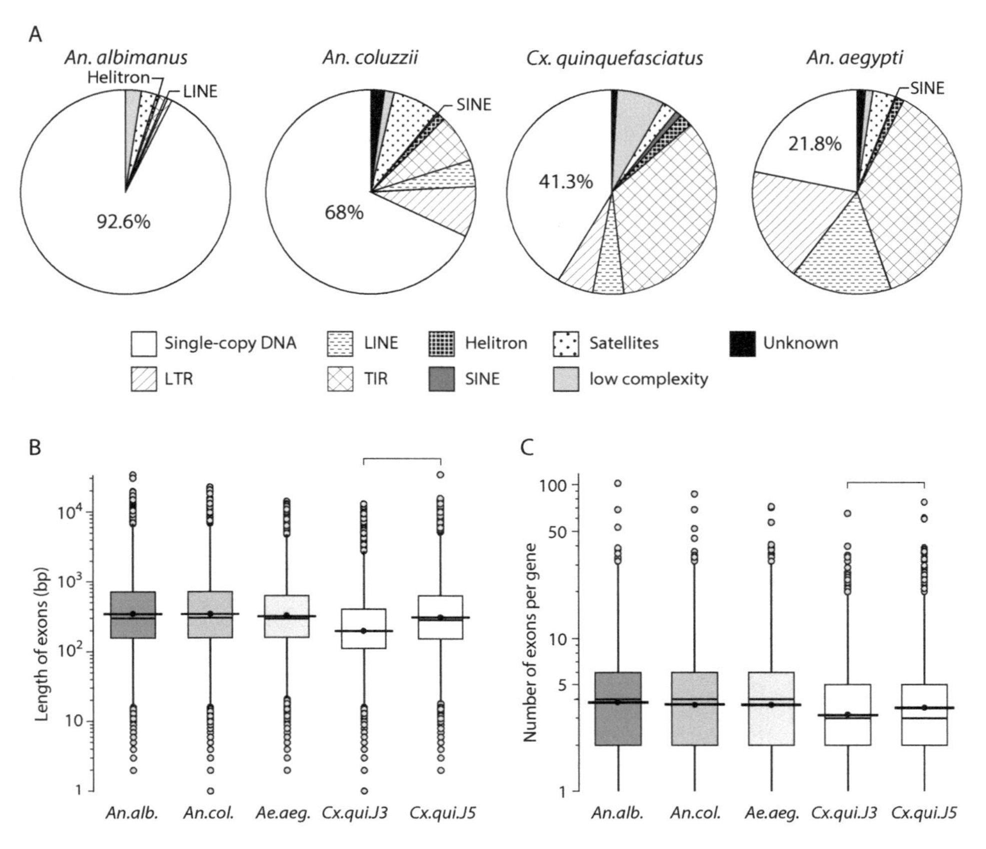

Figure 2.1 **A**. Genome composition of four mosquitoes (see also table 2.1). The relative abundances of different types of DNA sequences are indicated. LTR, LINE, TIR, Helitron, and SINE are different types of transposable elements. **B**. Lengths of exons in the genomes of four mosquitoes, and the numbers of exons per protein-coding gene. The average exon length is about 250 bp. **C**. Number of exons per protein-coding gene. Statistical analyses indicate significant heterogeneity among species. From Ryanzansky et al. (2024).

though they average about 400 amino acids across all species that were examined (figure 2.1B). The average number of exons (coding regions) in protein-coding genes is likewise quite uniform across species, being very close to four (figure 2.1C), although somewhat lower in *pipiens*.

This uniformity of the protein-coding fraction of genomes belies the differences in a myriad of biological properties, as will become clear in the following chapters. This means that the similar sets of genes must be functioning differently, in order to generate the diversity of phenotypes. It is widely thought that much of the diversity among organisms is not primarily due to having different genes, but rather in how these genes are used—in other words, where and when they are expressed, and in what abundance. This is similar to constructing a building, where an architect can use the same building blocks (proteins) to create a wide variety of structures. It is a matter of which and how many building blocks are used, where and at what point in the construction. This is not to say that the proteins are constant across species. Rather, it emphasizes that the major genetic underpinning of phenotypic variation is likely due to variations in gene expression.

Recombination

A basic property of genomes is recombination. Over the entire genome, the total recombination map of these mosquitoes is remarkably uniform, around 200 cM (centimorgans, or percent of chromatids that are recombined). When considering how this translates to recombination per length of DNA, however, there is considerable variation (last column in table 2.1). Among eukaryotes, a generality (often violated) is that there is about one cM per Mb of DNA. *An. gambiae* and *C. pipiens* conform to this, although the number for *pipiens* is a bit low. *Ae. aegypti* stands out as being particularly low in recombination per length of DNA. This could be related to its large numbers of TEs. In models of the evolution of TEs, it is generally assumed that most insertions of DNA are deleterious, more likely to disrupt genome function in a negative way than to improve it. So, teleologically, genomes are fighting to remove new insertions of TEs. The efficiency of removal of deleterious mutations is related to recombination: higher recombination leads to more efficient removal, because it reduces the linkage to parts of the genomes that are needed for survival (Kent et al. 2017). But this brings up the issue of whether low recombination is due to high level of TEs, or the high level of TEs is due to low recombination—a chicken or egg problem so often encountered in evolutionary considerations. Recombination rate is an important consideration in many aspects of mosquito research, especially in the topics covered in chapters 9 and 10.

Linkage Disequilibrium

Linkage disequilibrium (LD) is the nonrandom association of linked DNA sequences. Over many generations (at equilibrium), recombination should shuffle linked genes, so that eventually they become randomly linked. The various combinations of alleles at loci occurring together on the same chromosome (multilocus genotypes) should simply be the product of the frequencies of their occurrence in a population. When this is not the case, the loci are said to be in linkage disequilibrium.[4]

The level of LD in a population is determined by three major factors: the rate of recombination between loci, the age of a population, and population size. Obviously, the greater the number of recombinations, acting over longer time (number of generations), the less LD there will be. Population size also enters in, as the association of linked alleles is expected to be exactly that of the product of their frequencies, but only if the sample size is infinite. The size of a population determines the sample size of scrambled chromosomes produced by recombination in gametes from one generation to the next. Small populations (small sample size) produce nonrandom associations by sampling only a small fraction of the gametes each generation, which increases LD. So, while the rate of recombination may be a constant for a species, the other two factors (population age and size) are not. Another factor causing nonrandom associations of alleles at different loci is epistatic selection, where particular combinations of alleles (either linked or nonlinked) may confer high fitness. Thus *the level of LD is unique to each population*, so there is no single LD for a species.

Linkage effects and LD have become especially important in the era of large-scale whole genome sequencing, where tightly linked loci, and even adjacent nucleotide sites, are part of the data, and LD becomes apparent. In many analytical procedures in population genetics involving multiple loci, it is assumed that each locus varies independently, an assumption that is violated when LD is present. Selection on genome regions can be inferred by patterns of LD, as well as inferences on the demographic history of the populations that have been sampled. Levels of LD determine how successful and accurate genome-wide association studies (GWAS) are—a major method for understanding the genetics underlying complex phenotypic variation.

Aedes aegypti

LD in several populations of *aegypti* has been determined. The method of determining LD in whole genome data is to calculate the correlation (r^2) of nucleotide variants along a chromosome, which is expected to be zero if no LD exists. Figure 2.2A illustrates data from two sets of populations: one from Africa (Aaf) and one outside Africa (Aaa). It is clear that LD is greater in Aaa than Aaf, which is to be expected if Aaf is older than Aaa. Importantly, these data provide guidance for GWAS studies. Significant LD exists, up to about 1 kb in Aaf and 5 kb in Aaa. Thus, in order to have statistical power, sample sizes in GWAS studies need to be greater if Aaf is being studied than if Aaa is examined.

Anopheles gambiae

Figure 2.2B shows LD in five populations of *gambiae* s.s. and one of *coluzzii*. Again, sufficient LD exists (at least in some populations) to perform GWAS, and there is considerable variation across populations. The Kenya sample is particularly notable in having very high LD. This could be due to the population that was sampled (Kisumo) having been recently founded by relatively few individuals. Other data consistent with this interpretation show that this population has a deficit of rare alleles, longer runs of homozygosity, and a higher sharing of haplotypes among individuals (Anopheles Genome Consortium 2017).

Culex pipiens

A reliable genome assembly for *C. pipiens* has only recently been produced (Ryazansky et al. 2024), which is a prerequisite for estimating genome-wide LD. LD decays very rapidly in these populations of *pipiens*, essentially vanishing at about 200 bp (Y. Haba and C. McBride, personal communication). This does not bode well for GWAS work.

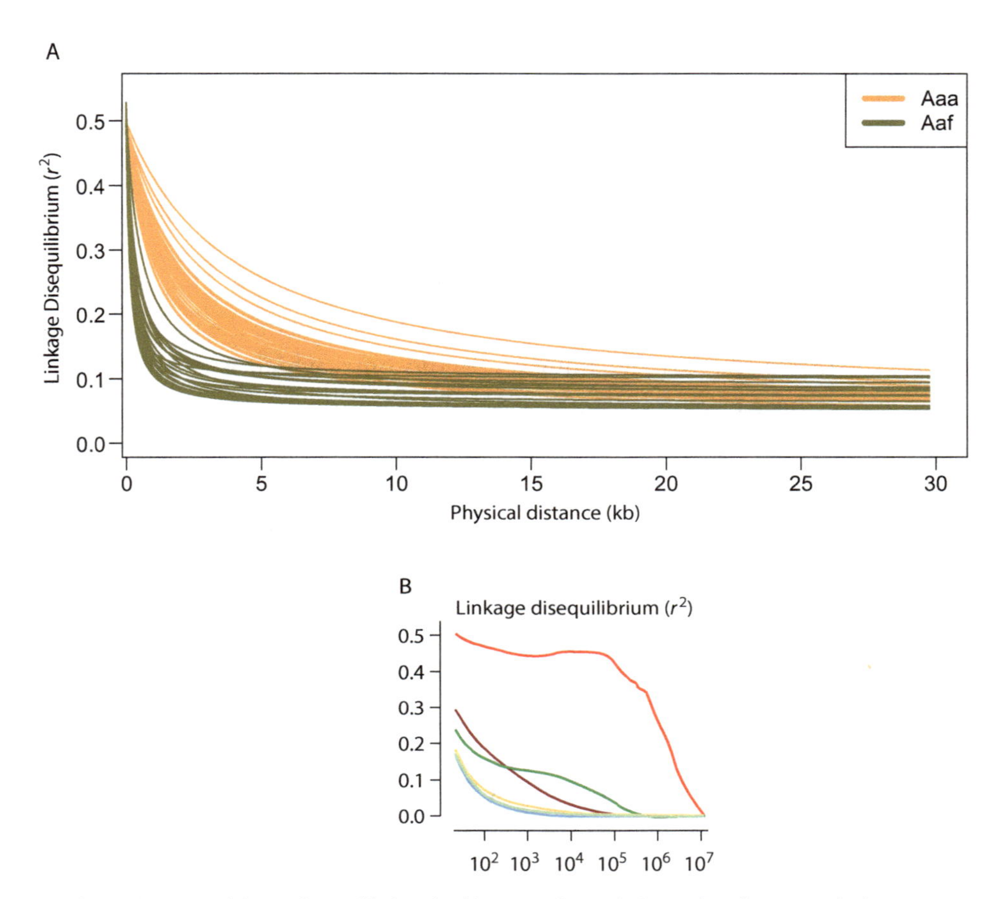

Figure 2.2 **A**. Linkage disequilibrium (LD) in several populations of *Aedes aegypti*. The upper lighter lines indicate Aaa outside Africa, and the lower darker lines are for Aaf from Africa. Here (and in **B**), the y-axis is the correlation coefficient r^2 between adjacent loci, separated by the physical length in kilobases (or, in **B**, the number of nucleotides) on the x-axis. From Crawford et al. (2024). **B**. LD in *Anopheles gambiae* complex populations. The three super-imposed lowest lines are *gambiae* s.s. from West Africa, the intermediate green line is for Gabon, and the intermediate dark-red line is for *coluzzii* from Angola. The upper line, showing considerable LD, is for *gambiae* s.s. from Kenya. From *Anopheles gambiae* 1000 Genomes Consortium (2017).

Karyotypes

While genome size varies among these three mosquitoes (table 2.1), all these mosquitoes have only three chromosomes (diploid number is six). Synteny[5] in the arms of these chromosomes has been conserved in

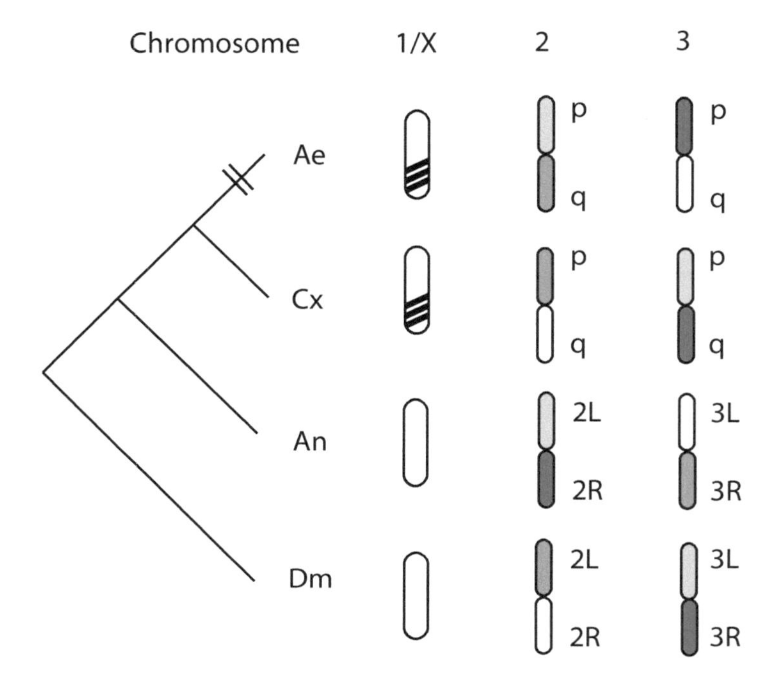

Figure 2.3 Synteny among chromosome arms of three mosquito genera and *Drosophila melanogaster* (Dm). Arms with the same shading are syntenic (contain the same sets of genes). The X chromosome is acrocentric in Dm and An (*Anopheles*), but submetacentric in Cx (*Culex*) and Ae (*Aedes*). Dashed lines on the 1/X chromosome indicate arm exchange. From Arensburger et al. (2010).

evolution, and homologies among them can be inferred (figure 2.3). Ryazansky et al. (2024) provide a further discussion of karyotype evolution among mosquitoes.

Karyotypic Differences

Two major differences occur among the three mosquitoes. First, *aegypti* and *pipiens* do not have a Y chromosome, whereas *gambiae* does. This means their sex determination is different (see the next section). Secondly, the ease in examining polytene chromosomes differs, meaning that the ease of detection of chromosomal variants, such as inversions, is not uniform across species.[6] *An. gambiae* has tissues from which usable polytene chromosomes can reliably be prepared, and a large body of literature on this group's inversion polymorphism is available. Figure 2.3 shows such chromosomes from the *gambiae* complex. While salivary glands are usually the tissue of choice to examine polytene chromosomes in Diptera, in *gambiae* the best preparations are made from nurse cells in gravid females. Chromosomal inversions play a crucial role in understanding the *gambiae* complex, as naturally occurring inversions are key in many aspects of this mosquito's biology (chapters 8, 9, and 10). *C. pipiens* polytenes can produce reasonable preparations, although not as easily as those from *gambiae*. It is virtually impossible to prepare readable polytene karyotypes with *aegypti*.

Sex Determination

While *gambiae* has a typical XY sex-determining system, *aegypti* and *pipiens* are unusual in this regard. In these two species, the smallest chromosome is designated chromosome 1, which is metacentric (two arms), with one arm homologous to the acrocentric X in *gambiae* (figure 2.4). Chromosome 1 in *aegypti* and *pipiens* has a sex-determining locus called the "M locus." Two alleles (M and m) exist, with heterozygous M/m being males and homozygous m/m being females. This locus is about 1.5 Mb in *aegypti* and contains two genes, *myo-sex* and *Nix* (B. Matthews et al. 2018). Aryan et al. (2020) have shown that it is the *Nix* gene that is crucial in determining sex, and *myo-sex* is needed for male flight, which is not surprising, given that this is a myosin gene.

Less is known about sex determination in the *pipiens* complex. It has a single locus (Gilchrist and Haldane 1947), presumably homologous to M/m in *aegypti*, that has been mapped to chromosome 1 (Mori et al. 1999). The *myo-sex* gene is found near the centromere on chromosome 1. No sequence homologous to the sex-determining locus *Nix*, however, was found in the best genome assembly of the complex, *quinquefasciatus* (Ryazansky et al. 2024). This gene was also found in 14 other diverse species of mosquitoes (Biedler et al. 2022).

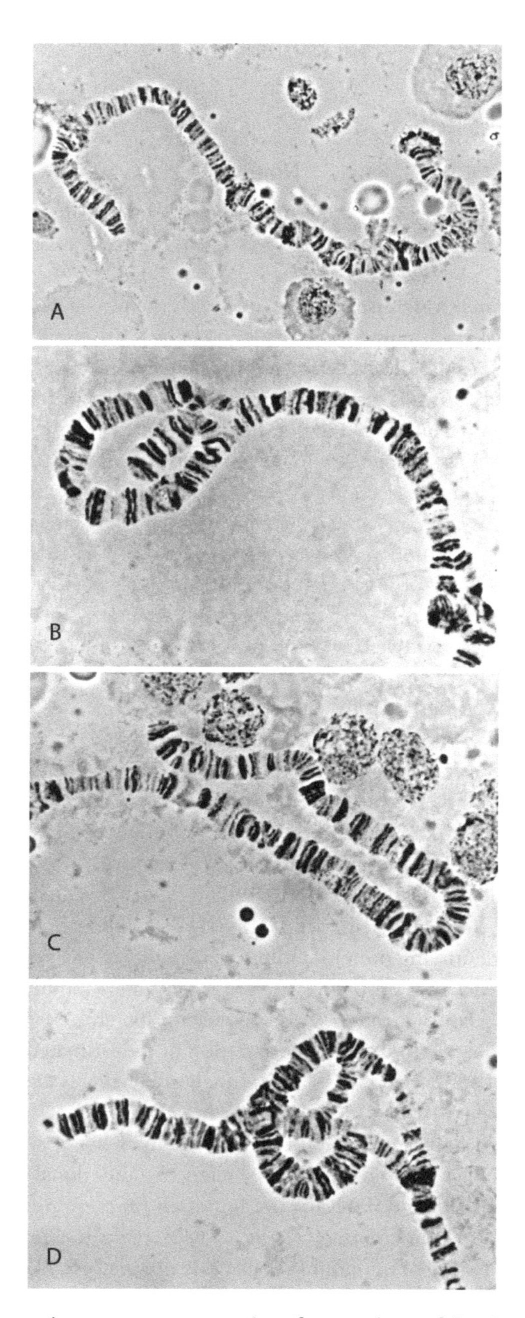

Figure 2.4 Polytene chromosome preparations for members of the *Anopheles gambiae* complex. **A.** *An. quadriannulatus*: 3R homozygous. **B.** *An. arabiensis*: 2R double nonoverlapping inversion heterozygote. The long near tip is 2Ra, and the short nearer centromere is 2Rc. **C.** *An. arabiensis*: 2R homozygous. **D.** *An. arabiensis*: single 3Ra inversion heterozygote. Photos provided by Vincenzo Petrarca.

Notes

1. To put these genome sizes in the context of other organisms, the human genome is ~2,900 Mb, and *Drosophila melanogaster*, ~220 Mb. Humans are estimated to have ~30,000 genes, and *Drosophila*, ~14,000.

2. There is an interesting history associated with the terms "satellite DNA" and "heterochromatin." When attempts were made to characterize genomes in the 1960s, a favored technique was to use ultracentrifugation, using cesium chloride (CsCl) gradients. DNA from an organism is mixed with CsCl and run at very high speed for 24–48 hours. Subjecting CsCl to prolonged high g-force centrifugation produces a density gradient of CsCl. DNA sequences vary in density, depending on the frequencies of A/T and G/C pairs, with fragments rich in A/T having a lower density than G/C-rich fragments. DNA fragments migrate to their equilibrium densities. The repeats in satellite DNA are 80 to 90% A+T compared to about 40 to 60% A+T in the rest of the genome. The result is one large band of DNA and one or a few small satellite bands at lower densities. These were eventually shown to consist of highly repetitive A+T-rich sequences.

In the early days, various dyes that bind to DNA were used to visualize chromosomes. The intensity of staining along chromosomes was seen to vary, with areas near centromeres being particularly darkly stained. These areas were called heterochromatin, while the rest of the chromosome was known as euchromatin.

The first use of chromosome *in situ* hybridization by Pardue and Gall (1970) demonstrated that the darkly stained regions near centromeres were satellite DNA. In a memorable demonstration of the stability of DNA, around 1990 I was visiting Joe Gall in his office at Yale University to get advice on performing *in situ* DNA hybridization. He calmly took a slide box sitting next to his microscope and pulled out the *original* 20-year-old slide and put it under the scope. It was beautiful!

3. Rigorously identifying protein-coding "genes" in a whole genome sequence (WGS) is difficult. It depends on the quality of the genome assembly and the definition of gene. The latter is a concept that has undergone considerable change as our understanding of genomes has evolved. Thus the variation in the estimated number of protein-coding genes among species may partly be due analytical artifacts and idiosyncrasies in how "genes" are defined.

4. The terms gene, allele, locus (plural loci), and markers are sometimes used interchangeably, although they subtly differ. "Gene" refers to a functional unit of the genome, a region that is transcribed, and the RNA product itself functions (e.g., tRNAs, rRNA, etc.) or is translated into a functioning protein. "Allele" refers to alternative sequences of a gene that differ because of mutations (nucleotide differences). "Locus" refers to a position in the genome. It can be applied to genes or to any unit of the genome—even a single nucleotide. How these terms are used is usually clear from the context. For example, two genes may be on the same chromosome, with alleles at these loci being nonrandomly associated (in LD). "Markers" is a general term that denotes experimentally observed, variable sites in a genome. These may be single nucleotides, whole proteins

(allozymes), or microsatellites (differences in the number of tandem repeat modules in a unique repetitive motif).

5. "Synteny" refers to the conservation of linkage of homologous genes on the same chromosome arm. The arms of these mosquitoes' chromosomes tend to have sets of genes that are conserved, so homology of these arms can be inferred.

6. Polytene chromosomes form when DNA replication occurs without cell division, known as endoreplication. Each homologous chromosome undergoes approximately eight DNA replications, producing 512 copies, with the replicated strands remaining paired. Homologous chromosomes pair in cells with polytene chromosomes, meaning that the microscopically detectable banding patterns represent 1,024 strands of DNA. Only the euchromatin undergoes endoreplication, which is largely the single-copy genome. Cells with polytene chromosomes are found in mosquitoes' salivary glands, nurse cells, and malphigian tubules. Once formed, the cell cannot divide.

CHAPTER 3

Mating and Oviposition

> A mosquito's hum may drive humans crazy, but to other mosquitoes it's love at first buzz.
>
> —*Science*, June 9, 2006

Mating

General Issues and Patterns

Since there are 3,700 formally described species of mosquitoes—and, doubtless, many more that remain to be described—in any given locality, tens or hundreds of different species coexist (are sympatric). When it's time to mate, males and females of the same species need to recognize one another to ensure the species' continuation. In these three mosquitoes, this is done through a variety of mechanisms.

The first step is for males and females to find one another. In the huge three-dimensional space of nature, these small insects are unlikely to randomly collide with a member of the opposite sex in their same species. Mosquitoes mostly use male swarming to attract females. Swarms of tens to thousands of males attract females that fly into the swarm, are grabbed midair, and often fall out of the swarm, *in copula*, onto the ground. Male swarming is common in insects (Downes 1969).

Male swarms are not a strict requirement, however, at least for *ae-*

gypti and *pipiens*. For example, in the *molestus* form of *pipiens* that breeds underground in a confined space, mating takes place in the absence of swarms. Single-pair matings (one male, one female) of *aegypti* can readily be made in small cages in the lab. When *gambiae* is first brought into the laboratory, male swarming remains a requirement, which makes laboratory rearing difficult, unless very large cages are available. Over generations of lab breeding, *gambiae* may adapt to mating in smaller spaces. To overcome the swarm requirement, *gambiae* can also be force mated (Bryant and Southgate 1978).

One important component of successful mating, especially in insects, is the compatibility of their genitalia. Male structures must be constructed to effectively transfer sperm to females. The compatibility of male and female genitalia (sometimes described as "lock and key") ensures the species-specific transfer of sperm. Male mosquitoes have more elaborate and variable genital structures than females. The most important are the claspers and aedeagus (penis). Claspers are used to grasp the female and position the aedeagus to enter the female's oviduct, thus allowing a sperm transfer. The evolution of male genital structures is generally faster than other body parts. Because of this, male genetalia are often used to identify and define species, while females are less morphologically distinct.[1]

Male mosquitoes have their genitals upside down upon eclosion (adult emergence from the pupal case), and it takes about 24 hours to rotate them to a usable position. This obviously delays male mating, although spermatogenesis has taken place in the pupal stage. Females are refractory to mating for about two days after eclosion. Speculatively, this delay in mating could be an adaptation to avoid inbreeding by reducing the probability of mating between close relatives eclosing from the same larval site.

In addition to sperm, males transfer several other molecules in the ejaculate (semen) that are synthesized in the male accessory glands, or MAG (Avilla et al. 2011). In *gambiae* s.s., an ecdysone (steroid hormone) transferred by males in their semen triggers oviposition, which is mediated by a specific response in females that is associated with the JNK signaling pathway (Peirce et al. 2020). Matrone is another transferred molecule, which renders a female refractory to further inseminations, making most females monogamous.[2] It has been shown, however, that in nature, females do sometimes mate multiple times. These studies have relied on genetic markers to infer the number of males that contribute to the offspring of a female collected in the field. The results are remarkably consistent across the three species, with multiple inseminations found at 6% in *aegypti* (Richardson et al. 2015), 2.5% in *gambiae* (Tripet et al. 2003), and ~5% for *pipiens* (Bullini et al. 1976). Thus it is safe to assume that monogamy holds at least 90% of the time. In artificial conditions, like partial enclosures in the field, multiple inseminations can be much higher (Helinski et al. 2012).

Given the prevalence of monogamy, how can a female produce fertile eggs throughout her life? Females have special organs for storing sperm after mating, called "spermathecae" (figure 3.1). They may have one (*gambiae*) to three (*aegypti*) such organs. Sperm stored in spermathecae remain capable of effective fertilization for weeks, thus relieving females of the need to engage with males again. Christophers (1960) cites studies indicating *aegypti* females can lay fertile eggs for 40–60 days after a single copulation.

Aedes aegypti

While initially thought not to swarm,[3] *aegypti* males congregate near sources of blood meals needed by females in what could be considered swarms (first clearly demonstrated in nature by Hartberg 1971). Carbon dioxide is a major attractant for females seeking a source of blood, and males have CO_2 receptors (Erdelyan et al. 2012). Given that males do not take blood meals, their attraction to CO_2 can only be explained by their desire to meet females at blood sources.[4] Once males and females are sufficiently close to one another, chemical and, especially, acoustic signals are used to complete their mating. Figure 3.2 summarizes steps in the courtship ritual.

Wingbeat frequencies for both males and females are especially important, and the use of this signal is complicated. Only a minority of attempted copulations result in successful mating (figure 3.2), despite all of the females in these observations being three- to seven-day-old

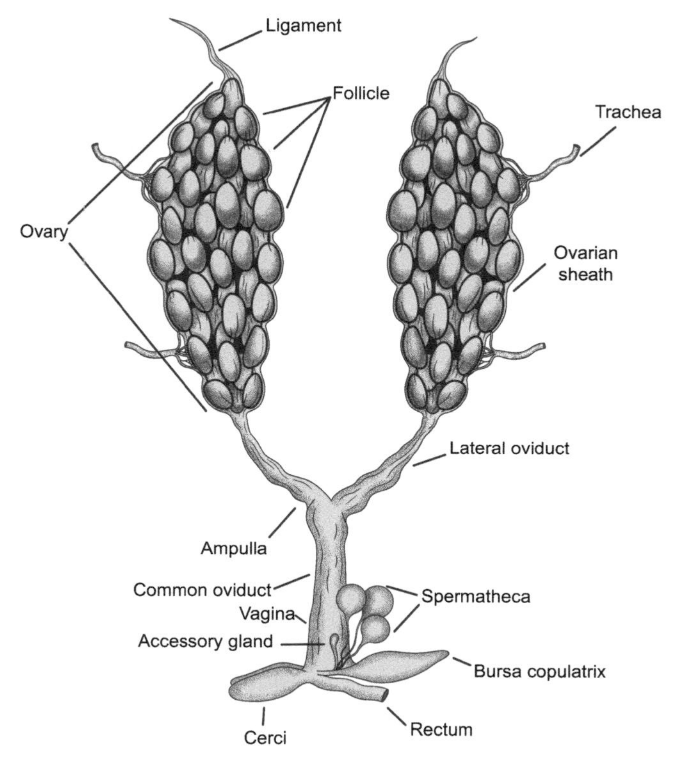

Figure 3.1 Internal reproductive organs of a female *Aedes aegypti*. Original drawing by Jacquelyn LaReau.

virgins. Cator et al. (2011) were able to record tone frequencies of wingbeats under field conditions and demonstrated that successful mating occurred when male and female wingbeats converged on a common frequency. Aldersley and Cator (2019) provide details of this phenomenon and follow-up work.

Anopheles gambiae

Members of this complex almost exclusively use male swarming to ensure that females and males congregate. Given the taxonomic and ecological diversity of this group, it is not surprising that it is difficult to generalize their mating behavior, although extensive research has been conducted.

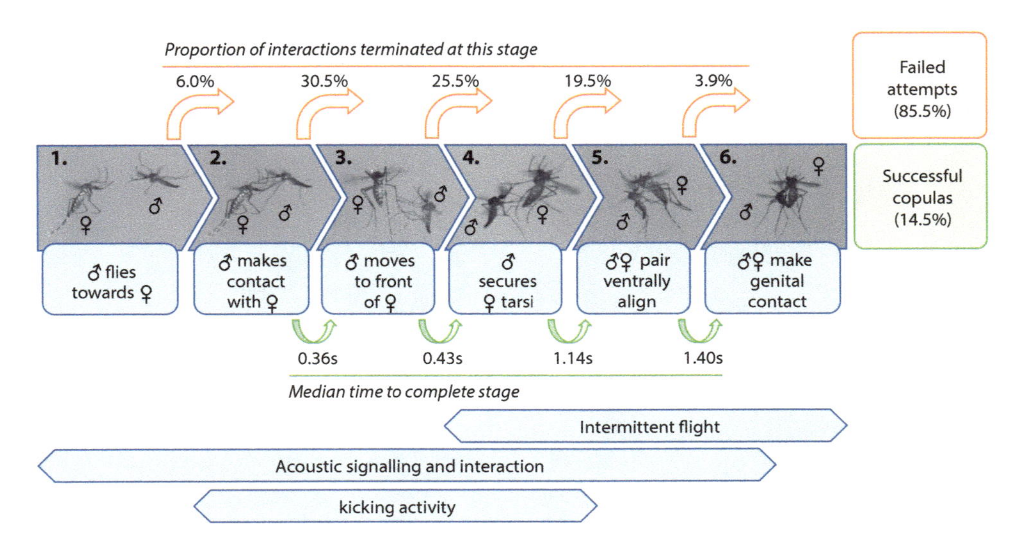

Figure 3.2 Schematic of mating stages in *Aedes aegypti*. From Aldersley and Cator (2019).

One of the more detailed studies is by Charlwood et al. (2002) for *gambiae* s.s. on the island of São Tomé off the west coast of Africa. They found male swarms that occurred two to three meters above markers (visual cues) with horizontal contrasts, such as grass and footpaths. Evening swarms occurred a few minutes before sunset and reached their maximum size within five minutes, with mating commencing seven minutes after their formation. Within the swarms, males were attracted to the sound of female wingbeats and did not respond to filter paper containing the odorants of squashed females. Up to 270 matings were observed in 20 minutes. The same genes that control circadian rhythms in insects, *per* and *tim*, are implicated in the timing of swarm formation and mating in *coluzzii* (G. Wang et al. 2021).

In Burkina Faso, *arabiensis* is reported to form mating swarms indoors as well as outdoors (Dabire et al. 2014). These authors also observed *arabiensis* swarming above larval breeding sites, which has not been reported for *gambiae* s.s.

The relationship between male swarm size and male mating success in *gambiae* s.s. was studied in Burkina Faso by Diabate et al. (2011). While male swarms varied in size between about 20 and 700, and larger swarms attracted more females, individual male mating success was independent of swarm size.

Because reproductive isolation among the various taxa of the *gambiae* complex largely occurs pre-mating, it is surprising that mixed male swarms can be found. For example, Dabire et al. (2013) observed that about 20% of swarms at a site in Burkina Faso were mixed M (*coluzzii*) and S (*gambiae* s.s.) forms. Despite the presence of more than one species of males in swarms, this does not seem to result in a breakdown in isolation. Out of 91 inseminated females isolated from mixed swarms, all contained conspecific sperm. A seemingly contradictory result was found by Niang et al. (2022) at the same site, as no mixed swarms were observed when using different DNA criteria to diagnose the species *gambiae* s.s. and *coluzzii*. In a study of the more distantly related siblings, *coluzzii* and *melas*, in Benin, Assogba et al. (2014) found no mixed swarms (out of 38 examined). Pombi et al. (2017) present a meta-analysis of a large number of studies addressing the possible role of mating isolation among the taxa of the complex.

Manoukis et al. (2009) studied individual male behavior in swarms of *gambiae* s.s. in Mali, using stereoscopic video imaging. They found swarms to be approximately spherical, with the highest density at the centroid.

Efforts to identify possible acoustic signals used in specific mate recognition in the *gambiae* complex have been largely negative. For example, no consistent wingbeat frequency differences were observed between *coluzzii* and *gambiae* (Simoes et al. 2017). Similarly, efforts to identify odor signals produced by cuticular hydrocarbons were largely negative, with only quantitative differences for a few chemicals found among taxa (Caputo et al. 2007).

Baeshen (2021) reviews swarming behavior in the *gambiae* complex, including a useful timeline of discoveries (with citations to key publications).

Culex pipiens

Like *gambiae* (but unlike *aegypti*), male swarming in *pipiens* is independent of blood meal hosts, and males seem to not be attracted to blood hosts (McIver et al. 1980). Given the wide distribution of *pipiens*, ranging from urban centers to natural woods and semideserts, it is not surprising that the visual cues used by males to form a swarm are variable and may constitute any prominent feature in the landscape. Referring to *pipiens* s.s., Knab (1906) stated that they "gathered over some prominent object such as a tree or a projecting branch, a bush, a corn-stalk or a person." Frohne (1964) observed that "trees provided the natural sites, but swarming also took place near buildings and telegraph poles and even at ground level in front of a flat slab of concrete topping a road bridge." In the laboratory, a 26 cm^2 piece of matte-black paper acted as a marker over which males swarmed (Gibson 1985).

Swarming occurs under low light conditions: dawn and dusk. In the eurogamous (open spaces) taxa *pipiens* and *quiquefasciatus*, Frohne (1964) noted that dawn swarms may be shorter (~30 minutes) than evening swarms (~60 minutes). Once males and females are sufficiently close, their mating ritual is not unlike that in *aegypti*, using acoustic and chemical signals, as well as physical interactions (e.g., kicking). Benelli (2018) summarizes some of this behavior when observed under laboratory conditions.

In stegnogamous (confined space) *C. molestus*, no swarming is noted, but there is a distinct mating ritual (figure 3.3). While it is initiated on surfaces, after a male mounts a female from behind a short flight ensues, during which copulation is accomplished. Copulation may be ventral to ventral or end to end (figure 3.3D and E).

Oviposition

After mating and taking a blood meal, it takes two to four days for a female to develop a fully mature set of eggs. Her behavior then goes into oviposition-seeking mode. During oviposition, the unfertilized egg travels from the ovaries through the oviduct, passing by the tubules from the spermathecae, which release a few sperm, one of which achieves fertilization (figure 3.1). The initial stages of all mosquitoes are aquatic, so eggs need to be laid on or very near water. In addition to suitable water (with correct salinity, temperature, odors, etc.), other environmental factors can affect oviposition. The amount of shade, as well as

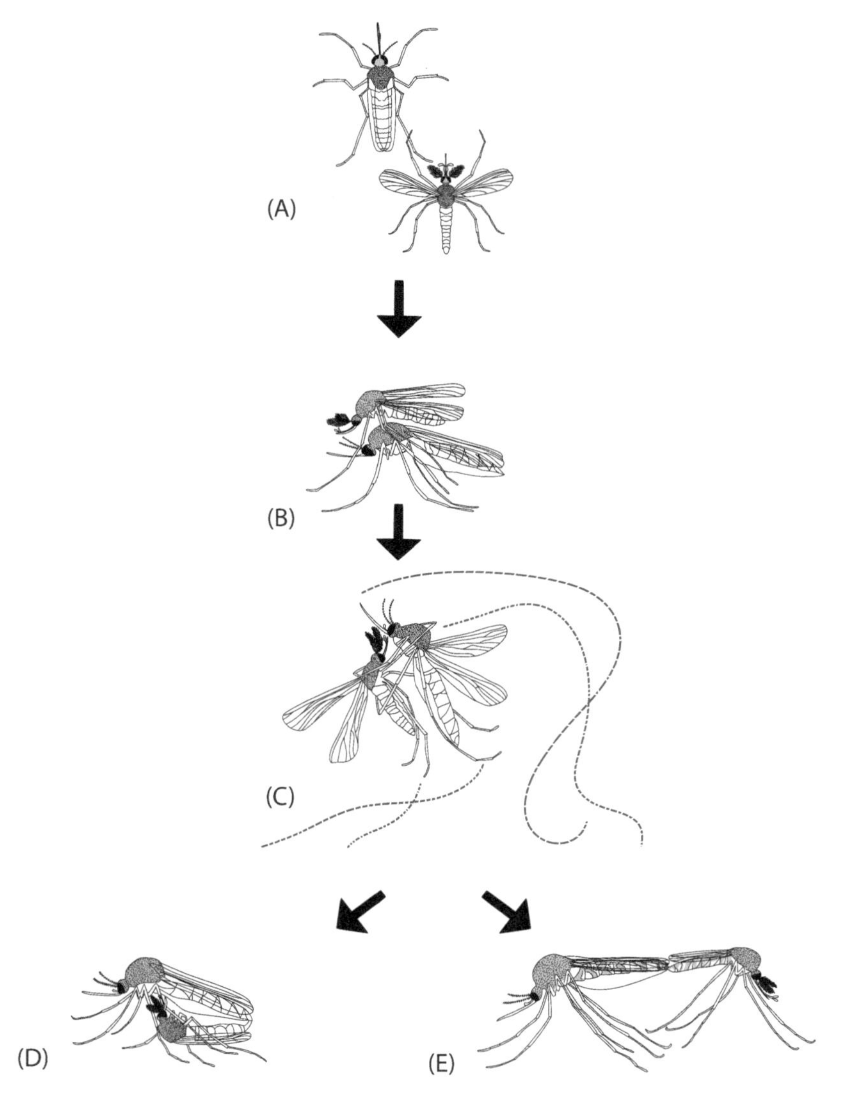

Figure 3.3 Schematic depiction of stages in the *molestus* form of *C. pipiens* mating in stenogamous (enclosed) conditions. **A**, tapping; **B**, mounting; **C**, co-flying; **D**, copulating ventral to dorsal; and **E**, copulating tail to tail. From Kim et al. (2018).

the color, volume, and movement of water (stagnant versus running) and the presence/absence of predators, can all affect female oviposition. Afify and Glaizia (2015) review more than 100 studies on mosquito oviposition behavior and can be consulted for entry into a large literature. Effects of predators on oviposition choices will be discussed in chapter 7.

Aedes aegypti

Ae. aegypti females lay eggs on moist, rough surfaces located just above the waterline of suitable freshwater pools. Presumably this is an adaptation to temporal shifts in the availability of water, which are common in this species' native habitats in Africa. In a wet season, new rainfall will raise the water level, flooding the eggs and stimulating hatching. Rising water makes it more likely that their development through to eclosing adults will occur before the water pool evaporates. If the water level does not initially rise to flood the eggs, such as during the dry season, the dormancy of *aegypti* eggs (3+ months) provides protection until this occurs. A rising water level makes it more likely the pool will remain long enough to complete larval development and adult eclosion.

The cues female *aegypti* use to choose an oviposition site have been studied in some detail, especially in the laboratory (reviewed in Bentley and Day 1989; Day 2016). The microbiome (the assemblage of microbes) in water is a primary factor in attracting mosquito oviposition. Autoclaved water is unattractive for oviposition. The chemical profile of compounds produced by bacteria and their effect on oviposition has implicated various carboxylic acids and methyl esters as being particularly stimulating for *aegypti* female oviposition (Ponnusamy et al. 2008). In addition to the visual and olfactory cues initially attracting gravid females, once a female who is ready to oviposit lands on a surface, she spends some time testing both the water surface and the side of the container by dipping the tip of her abdomen (ovipositor) onto the surface. She may also raise her tarsi to her antennae, in order to gain other cues. Insects are known to have chemosensory organs on the tip of their abdomens and tarsi. Videos of this behavior are available.[5] A specific gene and receptor, expressed in the legs and proboscis, that controls oviposition behavior in response to salinity has been identified in *aegypti* (B. Matthews et al. 2019).

In natural settings, direct observation of oviposition is difficult, so oviposition choices are inferred by the presence of *aegypti* larvae in the water. One generalization is that *aegypti* females avoid sunlit water pools for oviposition. Water in dark tree holes or well-shaded rock pools are used by sylvan Aaf. Even domestic Aaa avoid well-lit artificial pools. Chemosensory and olfactory cues are largely provided by the microbes living in a potential larval site.

One of the few studies (and the most detailed one) of oviposition of *aegypti* in nature is Xia et al. (2021). These workers examined the characteristics of potential oviposition sites in village-breeding and forest-breeding *aegypti* in Gabon and Kenya. Measured factors included physical characteristics (e.g., volume, shading, temperature, pH, etc.), volatile gas ("smell") above the water, and the microbial community in larval breeding sites in the field. Forest sites included tree holes and rock pools, and domestic sites included clay pots, discarded tires, and the like. Clear physical and microbial differences existed between potential oviposition sites in forests and domestic locales. Moving domestic containers to the forest, and forest mimics to the village, provided little evidence that forest-breeding and village-breeding females differed in their oviposition preferences. This supports the notion that *aegypti*, even in its native African habitat, is a generalist with regard to oviposition. This is consistent with observations of Aaa larvae outside Africa sometimes being found in natural sites like tree holes and leaf axles, and Aaf larvae in Africa sometimes found in human-generated containers in villages and cities (at least in the last 40 years).

Presence/absence of both conspecific larvae and other aquatic invertebrates can also affect female choices for oviposition. Generally, conspecific immature stages in water—either living larvae or shed exoskeletons—attract female *aegypti* to oviposit (Wong et al. 2011). On the other hand, a high density of larvae (overcrowding) or larvae infected with pathogens can repel oviposition by females (Zahiri and Rau 1998). A pheromone from larvae has been proposed as the signal for this behavior (Seenivasagan et al. 2009; see also Benzon et al. 1988). Studies on oviposition preferences with both *aegypti* and *albopictus* larvae revealed complex behaviors that are dependent on density (Gonzalez et al. 2016). Surprisingly, the presence of the larval predator *Toxorhynchites* may increase oviposition by *aegypti* females, evidently due to the increase in bacteria emanating from killed larvae (Albeny-Simoes et al. 2014).

An important aspect of *aegypti* oviposition is that females do not lay many eggs per site at any one time. This has been called "skip oviposition." It is an example of bet hedging—in this case, literally not putting all your eggs in one basket. Using more than one oviposition site reduces competition with siblings, as well as diminishes the chance of losing all offspring due to a pool drying up or invasion by a predator. Direct evidence of this in nature is provided by genetic studies of collections in individual larval breeding sites. In both Trinidad (Colton et al. 2003) and Puerto Rico (Apostol et al. 1994), skip oviposition was confirmed for *aegypti* under natural conditions. Because of this behavior, the distance between suitable oviposition sites can affect the dispersal behavior of female *aegypti*, as they disperse more when needing to fly further to find multiple sites for skip oviposition (Reiter 2007; see chapter 7 for more details). Harrington and Edman (2002), however, found indirect evidence contradicting skip oviposition in *aegypti*.

While female *aegypti* generally prefer relatively clean fresh water for oviposition, Barrera et al. (2008) have documented larvae in septic tanks in Puerto Rico. This is a habitat more generally occupied by members of the *pipiens* complex (see below). There is no evidence that *aegypti* living in septic tanks are genetically distinct from nearby surface populations (Somers et al. 2011). *Ae. aegypti* larvae have also been collected in brackish water in Sri Lanka (Ramasamy et al. 2014). There is little or no evidence of genetic differences between brackish water and freshwater collections in the same area (unpublished data from Etowah Adams et al.)

Anopheles gambiae s.l.

As with mating and many other traits, it is difficult to summarize oviposition behavior in the *gambiae* complex, although a few generalities are evident. Like all anophelines, members of this complex lay eggs on water surfaces with floats attached (figure 1.1). *An. gambiae* may also lay eggs in mud that is adjacent to pools, with newly hatch larvae having been observed crawling to open water (Miller et al. 2007). Giglioli (1965) made detailed observation of *melas* performing oviposition behavior in a mangrove forest in Gambia, where mud also appeared to be the preferred surface.

Unlike *aegypti*, most members of the *gambiae* complex prefer cloudy or turbid water (McCrae 1984). Another difference is that *gambiae* s.l. females generally prefer to lay eggs in sunny pools in open habitats, consistent with algae being a major source of nutrition for them (Merritt et al. 1992; Gimnig et al. 2002; Munga et al. 2005). Such pools may be very small (e.g., from hoofprints of cattle along the edge of larger pools) or large, such as flooded rice paddies. Fresh water is used by most *gambiae* taxa, but *melas* and *merus* use salty or brackish water. Eggs are not resistant to drying, do not display the extended delayed development seen in *aegypti*, and do not remain viable for more than about eight days after laying.

The issue of whether *gambiae* exhibit skip oviposition is less clear. Two studies using genetic markers to infer the number of females contributing to offspring in pools produced similar results. For *gambiae* s.s., Chen et al. (2006) found that about 57% of females had offspring in more than one pool. For *arabiensis*, Odero et al. (2019) estimate that "in excess of 50%" of females laid eggs in more than one pool. Such inferences regarding a female's offspring coming from more than one larval breeding site may not represent skip oviposition in the same sense as in *aegypti*. It is conceivable that during any given oviposition bout (generally occurring under low light conditions at dawn and dusk), a female lays all her eggs that are at the oviposition stage in one pool, while on subsequent days she uses another pool. It remains to be confirmed whether *gambiae* females use more than one pool during a single oviposition bout.

Work on physical cues attracting *gambiae* oviposition includes the role of moisture in the substrate surrounding a pool and visual contrast (Huang et al. 2007).

Considerable work has been done on finding chemical attractants used by anophelines to choose oviposition sites. Like most mosquitoes, members of the *gambiae* complex avoid sterile water, so microbes likely play a role in choosing where to oviposit. Herrera-Varela et al. (2014) showed that, in the field, *gambiae* s.l. females readily laid eggs in lakes with an infusion of soil that had aged for six days. A similar water treatment with an artificial nutrient source (commercial rabbit food) was avoided by females. Autoclaving decreased the attractiveness of soil-infused water.

Lindh et al. (2008) attempted to identify either specific bacteria or chemicals accounting for an attraction for oviposition by *gambiae*. Seventeen bacterial species, isolated either from adult midguts or larval breeding, were studied, six of which significantly attracted oviposition. Later, a specific chemical from soil-infused water, the complex alcohol cedrol, was identified as a specific oviposition attractant (Lindh et al. 2015). Maize pollen has also been implicated as attracting *arabiensis* oviposition (Wondwosen et al. 2017).

The presence of eggs or larvae in potential oviposition water can also affect *gambiae* egg laying. Conspecific larvae at low densities attracted oviposition by *gambiae* s.s. females, whereas a high density of larvae repelled oviposition (Munga et al. 2006; Sumba et al. 2008). Because these studies used egg counts to infer oviposition, Huang et al. (2018) suggest that this may be an artifact of fourth instar larvae cannibalizing eggs and early-stage larvae. This is consistent with Schoelitsz et al. (2020), who showed that while early instar larvae attracted oviposition, fourth instars were a repellent. They identified four chemicals emitted by different larval stages that could be the signals used by females to distinguish water with different larval stages.

There may also be region-specific differences in oviposition cues for *gambiae* s.s. (Ogbunugafor and Sumba 2008). Larval breeding water collected in Kenya and Tanzania attracted nonrandom oviposition by females from the same region where the water was collected.

Culex pipiens

Members of the *pipiens* complex lay egg rafts, consisting of 100 or more eggs that are held together and float on water surfaces (figure 1.1). Females generally lay one egg raft at a time and may, over a lifetime, lay up to five rafts. Generally, eggs hatch in 24–48 hours, although hatching may be delayed by low temperatures. Because virtually all developed eggs in a female are oviposited at once in a raft, there is no possibility for skip oviposition. Siblings develop together.

As is the case for virtually all mosquitoes, organic matter and microbes in water are important in attracting oviposition by *pipiens* s.l. females, although the level of foulness is often greater than that for most mosquitoes, with *pipiens* s.l. larvae found in sewers in cities and aboveground cesspools used to store animal feces and urine on farms. Unlike the other two mosquitoes, members of the *pipiens* complex sometimes lay eggs on moving water, where larvae have been observed to develop into adults (Calhoun et al. 2007; Chaves et al. 2009).

It is almost impossible to characterize *pipiens* as breeding solely in containers, open water, enclosed water, small/large pools, or clear/dirty water. In different places, all these types of standing (and sometimes moving) water are used. Temperature and water color have been studied in regard to *pipiens* oviposition preferences (Gillespie and Belton 1980; Dhileepan 1997; J. Li et al. 2009). Studies of water infused with various substances indicated that hay and grass infusions attracted more oviposition than manure and rabbit chow (Jackson et al. 2005). In general, it is safe to conclude that oviposition choices for *pipiens* complex females are very broad, much broader than for the other two mosquitoes.

Perhaps the most striking aspect of *pipiens* oviposition is the role of conspecific eggs in stimulating females to lay eggs. Females deposit an apical droplet on each egg in a raft, which acts as a pheromone (designated MOP, mosquito oviposition pheromone) for other females to oviposit (Osgood 1971; Bruno and Laurence 1979; Laurence and Pickett 1985). The chemical in the droplet was identified as a complex organic compound, (-)-(5R, 6S)-6-acetoxy-5-hexadecanolide (Mihou and Michaelakis 2010). Of note, MOP acts not as an attractant but only as a stimulant to oviposit. Greater numbers of eggs are laid on water with MOP, but females are not attracted more to traps with MOP when compared with water (Leal et al. 2008; Fytrou et al. 2022).

Notes

1. Christophers (1960), Harbach and Knight (1980), Jobling (1987), and Wilkerson et al. (2021) provide much more detail on the anatomy of mosquitoes, including male and female genitalia. The terminology and detail of such descriptions can be overwhelming. Wilkerson et al. have a 12-page glossary of hundreds of technical terms describing mosquito genitals, and Harbach and Knight's book runs to 415 pages! Detailed and truly lovely photographs of the morphology of *Culex* sex organs can be found in Klaus et al. (2002).

2. G. B. Craig (1967) was the first to show that female monogamy in mosquitoes is due to a substance transferred in sperm. He called this substance "matrone," as it converted a mate into a matron.

3. Christophers (1960) states, "The swarming of males so conspicuously evident in many . . . mosquitoes . . . is not a feature with *Aedes aegypti*."

4. Opposite sexes of many insects find each other at feeding sites. This has been described as "mating at the dinner table." In addition to congregating at blood sources, there is some indication that *aegypti* males use an aggregation pheromone to form swarms (Fawaz et al. 2014), although this is controversial and remains to be confirmed.

5. See https://youtu.be/SEErn3XL1RQ.

CHAPTER 4

Larvae and Pupae

> Most of the pollution in the water already is dead animal and plant matter and building debris, . . . good stagnant nursery for mosquitoes.
>
> —James Wright

All mosquitoes have aquatic larvae that molt three times to produce increasingly larger larvae, called "instars." Almost immediately after oviposition, larval development begins inside the egg capsule (shell). Full development of first instar larvae, ready for life outside the capsule, takes about three days. An egg will not hatch until the larva is fully developed. Eggs remain viable for another 4–10 days for *gambiae* and *pipiens*. In the case of *aegypti*, these encapsulated larvae may remain in a state of suspended development for three to four months.[1]

Development time from first instar larva to pupa may be as short as five days with optimal temperatures and nutrition. Factors affecting the rates of larval development are reviewed in Couret and Benedict (2014) and Zapletal et al. (2019). The relative time spent in each instar is presented in table 4.1. Molting (ecdysis) from one stage to the next is controlled by well-studied hormones, including juvenile hormone and ecdysone (Clements 1992, 1999, 2012). Cold and starvation can extend the larval period to two or more weeks. During this brief time, growth is quite remarkable. The weight of an egg is approximately 1 ug (micro-

Table 4.1 Time spent in immature stages (larvae and pupae) for two species

Stage	*quinquefasciatus*	*aegypti*
First instar	14%	19%
Second instar	13%	14%
Third instar	17%	17%
Fourth instar	33%	27%
Pupa	23%	23%

Note: These are averages over a range of temperatures; *gambiae* likely has similar averages. From Rueda et al. (1990), as compiled in Wilkerson et al. (2021).

gram), and that of a fourth instar larva, 4,000 ug (Lang et al. 1965; van Handel 1993). Under optimal conditions, this means a doubling of weight roughly every 12 hours! Fourth instar larvae, weighing 4 mg (milligrams), are heavier than the ~2 mg adult females and ~1 mg adult males. This growth occurs despite larvae being highly active, earning the colloquialism "wrigglers."

Figure 4.1 shows the morphology of the three mosquitoes' larvae. Anopheline larvae can easily be visually distinguished from culicines (*Aedes* and *Culex*) by their lack of a siphon, as well as their positioning with respect to the water's surface (see figure 1.1). *Aedes* and *Culex* differ most conspicuously by the lengths of their siphons, as well as by the tufts and brushes at their posterior. Taxonomic studies can be done on larvae, but adults are the preferred stage for identifying species. Molecular approaches to species identification, which are based on DNA sequences, or barcoding (Hebert and Gregory 2005), are easily applicable to mosquito larvae. A recent search revealed reports of about 1,400 mosquito species with mtDNA cytochrome oxidase I DNA sequences (www.barcodinglife.org). This has become the most common method to determine species at the larval stage.

Not surprisingly, given their remarkable rate of growth, ingesting nutrients is the primary occupation of larvae. Feeding occurs by filtering small particles suspended in the water, including microbes (e.g.,

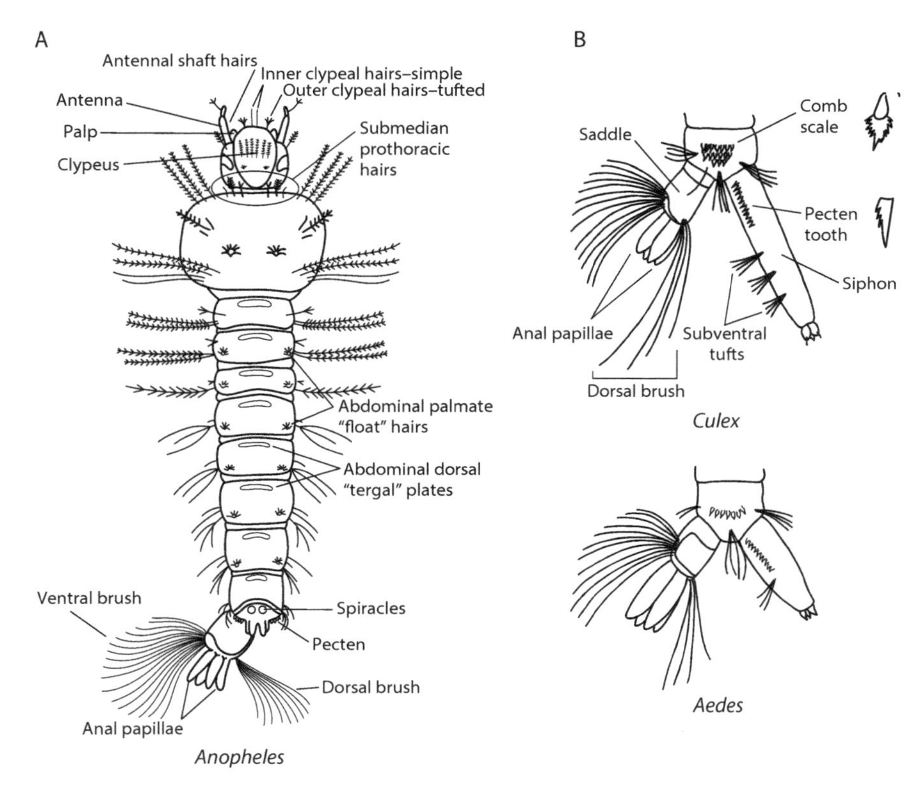

Figure 4.1 Basic anatomy of mosquito larvae. The major differences among the three genera are in their posterior terminal parts, with anophelines having no siphon (air tube), and *Aedes* having much reduced ventral brushes. Public domain online.

bacteria, fungi, and algae).[2] This may be done while hanging down from the surface (as depicted in figure 1.1), using brushes around their mouthparts to move water through a filter in the mouth. More-active feeding may occur when larvae dive to the bottom and use their mouthparts to scrape up and resuspend particles that are then ingested.

The microbiota at larval breeding sites, as well as the larvae themselves, have been characterized in a number of studies reviewed in Scolari et al. (2019), Strand (2018), Guegan et al. (2018), and Gao et al. (2020). These studies are largely descriptive as relatively little is known about how microbiota function and/or interact with mosquitoes. As might be expected, the microbiota in the water in which larvae are living affect larval gut microbiota, as well as adult gut microbiota. There is evidence that the microbiome can affect the efficiency of transmission of vector-borne pathogens (Dennison et al. 2014; van Tol and Dimopoulos 2016; Caragata et al. 2019), as well as insecticide resistance (Dada et al. 2018). Mosquitoes and microbiota are discussed further in chapter 11.

Larval Habitats

Aedes aegypti

Larval habitats of mosquitoes are dependent upon the oviposition choices of gravid females. As discussed in chapter 3, *aegypti* females primarily oviposit on the edges of still, relatively clean freshwater pools. These may be in natural containers (e.g., tree holes, rock holes, leaf axils, etc.) or human-generated containers (tires, tin cans, flowerpots, etc.). As a generalization, Aaf in Africa use natural containers, and Aaa outside Africa use artificial containers. There are many exceptions, however, and *aegypti* should be considered an opportunistic species, able to exploit a number of habitat types for larval development. Ouedraogo et al. (2022) recorded physical and chemical properties of larval habitats in indoor and outdoor sites in Ouagadougou, Burkina Faso, and they relate these factors to productivity and the size of eclosing adults. Xia et al. (2021) report similar data in studies of oviposition for sites in Gabon and Kenya (discussed in chapter 3).

Chadee et al. (1998) documented the opportunism of *aegypti* breeding sites in the Caribbean. Aaa larvae were reported in rock holes, calabashes, bromeliads, ground pools, coral rock holes, crab holes, and conch shells. Perhaps because Caribbean islands have fewer species of mosquitoes than most tropical regions, these unusual larval sites for Aaa are open niches usually occupied by more-specialized species.

There are also exceptions to the "clean fresh water" rule: septic tanks in Puerto Rico (Barrera et al. 2008) and brackish water in Sri Lanka (Ramasamy et al. 2014). Chadee and Martinez (2016) proposed that underground larval breeding in *aegypti* (in storm drains as well as septic tanks) is a response to climate change and, possibly, insecticide use.

Anopheles gambiae s.l.

This complex has a broad range of larval breeding habitats that differ among the named taxa, which include hoofprints made by cattle at the edge of stagnant water, rice fields, and brackish water in coastal areas.

The larval habitats of three members of the *gambiae* complex are quite straightforward. *An. bwambae* larvae are only found in geothermal hot springs in the Simliki Forest of Uganda (G. White 1985). *An. merus* and *melas* larvae are found in salt water or brackish water along the east and west coasts of Africa, respectively. While *melas* is restricted to coastal areas flooded by ocean water (Coetzee et al. 2000), *merus* can be found considerably farther inland and may occasionally even breed in fresh water (Kloke 1997; Bartilol et al. 2021). B. White et al. (2011) note that when *merus* is observed in places away from the coast, it is found in fresh water sites where *gambiae* s.s. is absent, suggesting competitive exclusion. *An. melas* larvae have been found in water that is up to 72% seawater, typically containing 3.5% salt (Bogh et al. 2003), and *merus* larvae in up to 200% seawater (Bartilol et al. 2021 and references therein). Bogh et al. (2003) also report finding *gambiae* s.s. in up to 30% seawater. B. White et al. (2016) document the comparative survival of members of the *gambiae* complex at different levels of salinity.

While *quadriannulatus* was originally considered a single species, it is now thought to be two species: *quadriannulatus* in the southern part of Africa, and *amharicus* in Ethiopia (Coetzee et al. 2013). Both species are found in close association with domestic animals, but little is known about their larval habitats.

While most taxa are reasonably specific to certain larval sites, *gambiae* and *arabiensis* larvae are often found together in much of sub-Saharan Africa, where the two co-occur (figure 1.9).

Photosynthetic bacteria and algae are dominant in *gambiae* larval sites (Kaufman et al. 2006; Mala and Irungu 2011; Y. Wang et al. 2011). This is consistent with members of this complex breeding in sunlit pools, as opposed to most mosquitoes, which use shaded larval habitats.

The preferred larval habitats of *gambiae* s.s. are temporary rain-dependent pools or puddles in the sun, restricting their breeding to the rainy season (Simard et al. 2009). Adult *gambiae* s.s. are not found in the dry season. These populations survive the dry season by aestivation *in situ*, rather than being re-founded by immigrants (Lehmann et al. 2010).[3] The larvae of *coluzzii* are most often found in stable, often human-generated water, such as irrigated agricultural fields and artificial lakes, and this species is reproductively active year round (Touré et al. 1994, 1998b). Lehmann and Diabate (2008) discuss in some detail the importance of permanent versus temporary water sites in the evolution of the diversity of the *gambiae* complex (see also chapter 10).

Edillo et al. (2006) studied physical properties of larval breeding sites at a locality in Mali where *gambiae* s.s. (S form in that paper), *coluzzii* (M form), and *arabiensis* coexist. Conductivity and the total amount of dissolved solids had the greatest effect on niche partitioning.

Culex pipiens s.l.

Members of this complex occupy the broadest (and foulest) larval niches. Vinogradova (2000) lists 11 types of larval habitats, ranging from flowing streams; swamps; shallow temporary pools; natural and artificial containers; and above- and belowground, relatively clean (as well as highly polluted) human-generated aquatic waste. While both *aegypti* and *pipiens* can breed in storm drains, *pipiens* s.l. can also breed in sewers containing human waste.[4] Indeed, females are attracted to manure-infused water for oviposition (Jackson and Paulson 2006).

Noori et al. (2015) characterized the effect of ionic concentrations on *quinquefasciatus* larval development. Curiously, larvae of members of the *pipiens* complex are not reported from salt water or brackish water. In the laboratory, 50% of *pipiens* s.s. larvae survived 28% seawater, but none at 34% (Jobling 1938).

Predators and Cannibalism

Mosquito larvae and pupae live in environments that have various predators, as well as sometimes being subject to ingestion by larger larvae of their own species (cannibalism). While predation cuts across all mosquito species, it is less clear how widespread cannibalism is in the three species of concern here. Quiroz-Martinez and Rodriguez-Castro (2007) provide a comprehensive historical review of studies on insect predators of mosquito larvae.

The most common predators are larvae of other insects, such as members of the order Odonata (Saha et al. 2012). Vertebrates (such as tadpoles) and other amphibians prey on mosquito larvae (DuRant and Hopkins 2008). Chandra et al. (2008) review 13 species of fish that have been tested for their level of mosquito larval predation and potential use in mosquito control. Copepods may also prey on mosquito larvae (Marten and Reid 2007). Among our three species, all are probably subject to larval predation, although not under all conditions. For example, Aaa that breed in temporary water sites in human-generated containers probably encounter fewer predators than *coluzzii*, which breed in larger, more permanent bodies of water. Service (1977) has suggested that predation is a major cause of mortality in *arabiensis* in the field. Few predators live in cesspools or sewers occupied by *pipiens*. Dambach (2020) and Roux and Robert (2019) provided recent reviews of mosquito larval predators.

The impact of larval predators on mosquito populations was clearly demonstrated by the now-classic work of Christie (1958). Semipermanent pools were constructed in Tanzania, into which a known number of first instar *gambiae* s.l. were introduced. The survival rate to fourth instar larvae was only 8%; removal of all non-mosquito fauna except notonectid bugs (backswimmers) resulted in 19% survival; and removal of these bugs led to 58% survival.

Mosquito larvae are known to be both predatory and cannibalistic, with the genus *Toxorhynchites* being the most common predator. It is unclear how common predation or cannibalism by our three mosquitoes is, especially under natural conditions. In chapter 3, possible egg cannibalism in *gambiae* s.s. was mentioned (Huang et al. 2018). Larval cannibalism in *gambiae* has been documented in the laboratory (Koenraadt and Takken 2003; Koenraadt et al. 2004), including intra-instar cannibalism (Porretta et al. 2016). Most often, cannibalism involves later (larger) instars eating smaller larvae.

A related issue is whether mosquito larvae can sense predators or potential cannibals and modify their behavior to avoid being eaten. The presence of conspecific larvae and crowding can change larval behavior, but it is not clear if this is related to potential cannibalism. Chandrasegaran et al. (2018) studied the length of wiggle bursts by *aegypti* larvae. Evidently fish (in this case, guppies) detect and consume larvae during these bursts of movement. When guppies were present in the water, larvae displayed shortened wiggle bouts. Similarly, both *pipiens* and *aegypti* larvae were observed to be less mobile in the presence of notonectid bugs, and some evidence indicates that a chemical cue may be involved (Sih 1986).

Pupae

About midway through the fourth larval instar, juvenile hormone disappears, larvae stop eating, larval structures begin to break down, and adult structures form. By the end of the fourth instar, adult structures (compound eyes and legs) are visible through the cuticle. This is called

Figure 4.2 Time-lapse photograph of *Anopheles gambiae* adult eclosion in the laboratory. Photograph by James Gathany, Centers for Disease Control and Prevention.

the "prepupa" stage. Upon pupation, initially the cuticle is very light colored, but it darkens over a timespan of a few hours. Pupation requires a sufficient accumulation of nutrients to tear down old structures and build new ones. Males, being smaller than females, require less in the way of resources and generally pupate first. Pupae do not feed and are relatively quiescent, usually hanging down on the water surface, breathing through a trumpet organ. After about two days, the adult ecloses (figure 4.2).[5] As is clearly seen in this figure, upon eclosion, an adult is very light colored. A few hours are required to obtain a dark, melanized cuticle. There are subtle differences in morphology among pupae for our three mosquitoes (see figure 1.1).

A practical aspect in mosquito pupae is the size difference in males

Figure 4.3 Sex-sorting device with *Aedes aegypti* pupae. Two glass sheets are held together to form a wedge, with the space between them decreasing from top to bottom. Smaller males form the lower row, with larger females above. Note the virtual absence of an overlap in sizes.

and females. Sexes can be separated at this immature stage, ensuring virginity when controlled crossing is to be performed. While this differentiation is fairly easy to distinguish visually, mechanical devices have been developed to separate the sexes en masse. The most common is a wedge made from two sheets of glass, through which water containing pupae is poured (figure 4.3).

Notes

1. This state of suspended development for mosquito eggs is sometimes—inaccurately—called diapause. Diapause in insects usually occurs in adults to survive the winter (chapter 1, note 5). Among our three mosquitoes, *pipiens* s.s. females enter diapause when stimulated by a combination of shortening day length and dropping temperatures. In the case of *gambiae*, aridity can induce diapause or aestivation in adults, allowing them to survive the dry season (chapter 7 and note 3 below). This is quite distinct from what an *aegypti* egg does, as no environmental cues are involved. While not accepted by all entomologists, I prefer to reserve the term "diapause" for an environmentally induced physiological reduction in metabolic activity, generally in adults. The term "suspended development" seems more appropriate for *aegypti* eggs.

2. Pollen produced by corn (maize) has also been suggested as a possible source of nutrients for *arabiensis* (Ye-Ebiyo et al. 2000), consistent with evidence indicating that maize pollen may stimulate oviposition by this species (see chapter 3). It is unclear how widely pollen may be ingested by mosquito larvae, but pollen is known to have a high energy content.

3. Aestivation is the state of reduced metabolic activity and cessation of reproduction that allows insects to survive for prolonged dry/hot periods. Physiological changes occur in response to increasing temperatures and aridity. *An. gambiae* adults may live up to seven times longer in this state (Omer and Cloudsley-Thompson 1970). Aestivation can be thought of as the complement to diapause, used to survive cold winters.

4. In some countries, rainfall runoff from streets is collected in underground systems (storm drains). Human waste from homes is collected in a separate system (sewers). The water in storm drains is obviously much cleaner than that in sewers, and these two human-generated water systems are distinct niches with regard to mosquito larvae.

5. Thanks to the efforts of wildlife photographer Steve Downer, a wonderful video of adult eclosion (not strictly "hatching"), *Male* Culex pipiens *Mosquito Hatching—UHD 4K*, is available online. If this video does not win over skeptical readers who remain recalcitrant to the notion that mosquitoes are elegant, then the task is hopeless.

CHAPTER 5

Adults

> Changing from a player to a coach, I felt like a mosquito in a nudist colony. I didn't know where to begin.
>
> —Pee Wee Reese

I begin with some general aspects of adult mosquitoes.

- Upon eclosion, the cuticle is quite lightly colored and soft (see figure 4.2). Very shortly thereafter, an adult expands its folded wings and is able to fly a short distance, where it then remains quiescent for several hours as the cuticle hardens and darkens.
- Males have their genitalia upside down upon eclosion and rotate them within 24 hours.
- The sexes can be distinguished visually. Males have much bushier antennae and a blunter abdominal tip (see figures 1.1, 1.2, and 5.1).

The remainder of this chapter will deal with dispersal, longevity, and adult feeding (excluding female blood meals, which is the subject of chapter 6).

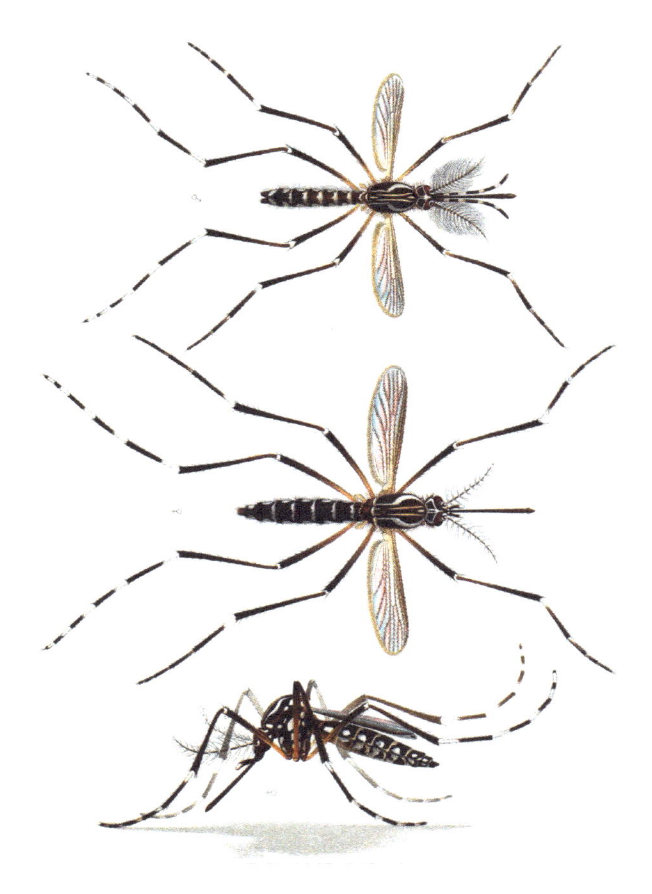

Figure 5.1 Prints of *Aedes aegypti*. Male (**top**), female (**center**), and side view of male (**bottom**). From Goeldi (1905).[1]

Adult Dispersion

Movement behavior of adult mosquitoes is a fundamental issue affecting many aspects of their biology, including determining distributions, the area over which females may spread diseases, population genetics parameters (such as effective population sizes), and guides to release strategies in attempts to control disease transmission. When considering the worldwide history of these three species (chapter 1), the importance of passive transport, mediated by humans, was emphasized. Here I will address active dispersal, or the movement of adults under their own power. Each species has a unique dispersal behavior, so this topic will be treated separately. Passive transport, including by wind, will be taken up in chapter 7.

Aedes aegypti

Several mark-recapture studies have been done on adult Aaa in cities or other human-associated habitats outside Africa, and one in Kenya, using introduced Aaa. Table 5.1 is a partial listing of such studies. They are heterogeneous in terms of methods: marking with dusts, using radioactive isotopes, reporting males and females separately, trapping by various methods, and the like. Despite this heterogeneity in methods and locations, these studies are reasonably consistent in showing that active movement by adult *aegypti* is quite limited.

One important variable for *aegypti* adult dispersal, especially for females, is the density of oviposition/larval breeding sites. The longest distances are reported for relatively dry localities in California (Marcantonio et al. 2019) and Texas (Juarez et al. 2020). Evidently, in order

Table 5.1 Summary of *aegypti* mark-recapture dispersal studies

Locality	Mean ♀♀	Mean ♂♂	Max ♀♀	Max ♂♂	Reference
Eastern Kenya	57	44	154	113	Trpis and Hausermann (1986)
Australia	56	35	160	160	Muir and Kay (1998)
California	220	240			Marcantonio et al. (2019)
Rio de Janeiro					Maciel-de-Freitas et al. (2007a)
suburb	85		250–350		
favela	40–50		100–150		
Hainan, China	10–40[1]	30–50[1]			Tsuda et al. (2001)
Thailand (3 sites)	90[2]		264[2]		Harrington et al. (2005)
Puerto Rico			102[2]		Harrington et al. (2005)
Texas	167/201[3]	242			Juarez et al. (2020)

Note: Numbers are estimates of meters traveled per day, except where noted.

1. Over more than one day, most often a lifetime.
2. Males and females were reported together. Mean distances are for between four and eight days after release.
3. Gravid/unfed females.

to skip oviposit (see chapter 3), females need to travel farther to find multiple suitable sites when such locales are more widely spaced. *Ae. aegypti* in a densely populated favela in Rio de Janeiro display less movement than in a suburb. Non–blood fed females moved more than gravid females in the Texas study (Juarez et al. 2020). In Harrington et al. (2005), both indoor and outdoor release sites were used, with no significant differences found.

In addition to mark-recapture, dispersal can be inferred by examining the relatedness of individuals across space. Jasper et al. (2019) studied a city in Malaysia and obtained results consistent with the limited dispersal revealed by mark-recapture. They found that the median distance from parent to offspring was 38 meters, and the neighborhood-area standard deviation was 91 meters. In other words, 95% of individuals are expected to travel 182 meters or less during their lifetime.

While most studies have supported the view that *aegypti* has a limited dispersal behavior, Reiter (2007) discusses cases documenting longer dispersal for this species. The effect of oviposition site spacing on female movements was discussed above and further discussed in Reiter et al. (1995) and Edman et al. (1998). Since most mark-recapture studies have been done in cities or other human habitats outside Africa, it is unclear whether occasional reports of longer-distance movements of adult *aegypti* are due to active dispersal or human-associated passive transport. For example, T. Schmidt et al. (2018) report full siblings found in ovitraps more than 1.3 km apart. If released far from their preferred habitat, adults may fly considerable distances, such as those released from a boat 900 meters offshore (Shannon and Davis 1930).

In cities in Australia and Trinidad, roads or highways have been shown to be barriers to active adult dispersal (Hemme et al. 2010; T. Schmidt et al. 2018).

All of these studies have been done on Aaa and thus are closely associated with human-disturbed environments. Little is known about the dispersal behavior of Aaf in its native African range. This also holds true for populations recently entering human habitats in Africa, for which very little is known about their basic adult behaviors (Facchinelli et al. 2023).

Table 5.2 Estimates of dispersal for *gambiae*

Locality	Taxon	Findings	Reference
Tanzania	*gambiae* s.l.	females 1 km, males 800 m in 23 days; max 3 km	Gillies (1961)
Burkina Faso	*gambiae* s.l.[1]	350–650 m/day, depending on the model	Costantini et al. (1996)
Kenya	*gambiae* s.l.	200–500 m after 2 days	Midega et al. (2007)
Gambia	*gambiae* s.l.	350 m/day, 10% >1.7 km	Thomas et al. (2013)
Burkina Faso	*coluzzii*	40–549 m over 7 days	Epopa et al. (2017)

1. *gambiae* s.s. and *arabiensis* were present in about a 2:3 ratio.

Anopheles gambiae

Mark-recapture studies on the *gambiae* complex have been performed in a number of localities in Africa (table 5.2). These are quite consistent in showing *gambiae* s.l. has higher degree of dispersion than *aegypti*. This research was done where *gambiae* s.s., *coluzzii*, and/or *arabiensis* are the predominant species. Other members of the complex may differ. For example, *melas* was documented to fly 1.6–4 km from its breeding sites in mangrove swamps to feed on humans in a village, and then returned to the swamp to deposit eggs (Giglioli 1965).

Culex pipiens

Like the other two species, several mark-recapture studies have been done on members of the *pipiens* complex (table 5.3). The methods and environments were quite heterogeneous, but their results have been fairly consistent, indicating that these mosquitoes disperse over relatively long distances (on the order of kilometers) over a lifetime. The Burkina Faso work (Subra 1972) is particularly interesting, as it documents the importance of environmental variables in determining how far these mosquitoes will fly. Perhaps not surprisingly, in rural environments adult mosquitoes travel greater distances, compared with urban settings.

One study is at odds with the others. Jones et al. (2012) speculate that the low dispersal distance they observed is due to the density of both favorable larval habitats and a high bird population (a primary

Table 5.3 Mark-recapture experiments on dispersal of *Culex pipiens* s.l.

Locality	Taxon	Findings	Reference
Japan, small island	*pallens*	500 m in 4 days, max 1.2 km	Tsuda et al. (2008)
Chicago	*pipiens*	mean 1.15 km, max 2.48 km	Hamer et al. (2014)
Hawaii, rainforest	*quinquefasciatus*	0.8–1.9 km in 11 days	Lapointe (2008)
California	*quinquefasciatus*	180 m/day, max 12.6 km	Reisen et al. (1992)
Rangoon	*quinquefasciatus*	max 0.4–0.7 km in 12 hours	Lindquist et al. (1967)
Burkina Faso	*quinquefasciatus*	urban 0.5–1.0 km in 12 days; rural 2.0–3.5 km in 12 days	Subra (1972)
Maryland	*pipiens*	<10 m/day	Jones et al. (2012)

blood source) in a suburb of Washington, DC, where the study was done.

Summary

It is important to emphasize again that studies of dispersal are dependent on environment and the research method used, which have been heterogeneous in mosquito studies. Nevertheless, it is quite clear that among our three mosquitoes, *aegypti* is quite sedentary, while members of the *gambiae* complex disperse considerably farther, and *pipiens* distances are intermediate. In this chapter I have considered just these subsets of the three genera, and a review of studies across all species in the genera produced similar conclusions (figure 5.2).

Longevity

While a number of studies on adult longevity have been done under laboratory conditions, I will confine my considerations here to work done on natural populations. A byproduct of mark-recapture studies done with recaptures over multiple days is that survivorship of the released adults can be estimated, so the mean number of days of life can be calculated. The longevity of adult mosquitoes plays a large role in disease transmission, as it takes several days for a pathogen ingested in

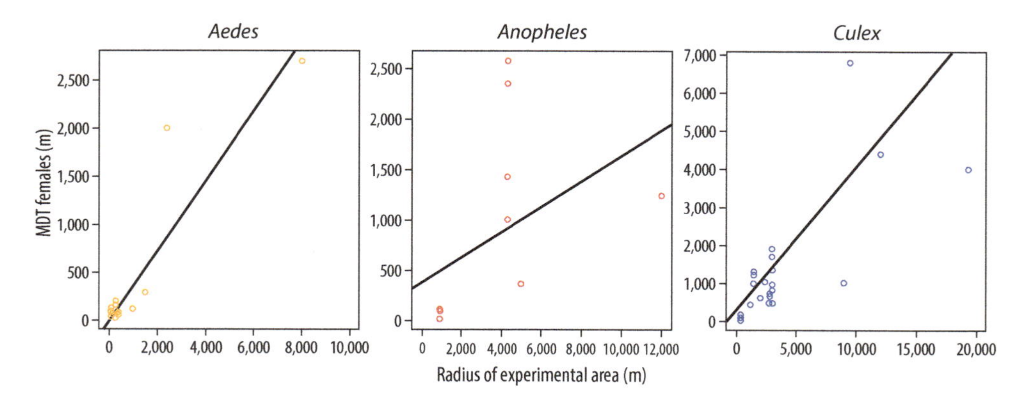

Figure 5.2 Estimates of dispersal from 774 mark-release-recapture studies performed on species of the three genera. Note the different x-axis scales. From Guerra et al. (2014).

one blood meal to arrive in the salivary glands in sufficient numbers to be transmitted in a second bite (chapter 11).

Aedes aegypti

Table 5.4 shows estimates of the daily survival of adult *aegypti* from release-recapture experiments (see also figure 7.1). Results are notably consistent across studies: males have daily survival rates between 70% and 80%, while the rates for females are between 80% and 90. The estimated median number of days they have survived, as well as the number of days for 10% survival, assumes a constant survival rate. Based on releases of laboratory-reared adults of various ages, however, this assumption is violated, as *aegypti* adults do "age," and older adults die at a higher rate than younger ones (Harrington et al. 2008).

Like the dispersal studies mentioned above, all of these studies were done on what would be considered Aaa in human-disturbed habitats. Survival rates of Aaf adults in native habitats in Africa are unknown.

Anopheles gambiae

Few estimates of the survival/longevity of adults have been made for this complex. Gillies (1961) reported a "mortality" of 16% per day, resulting from a combination of death and emigration. Midega et al.

Table 5.4 Daily survival estimates for natural populations of *aegypti*

Locality	Females	Males	50%	10%	Reference
Thailand	0.81	0.70	3.3/1.9	10.9/6.5	Sheppard et al. (1969)
	0.88[1]		5.4	18	
Kenya, village	0.89	0.77	5.9/2.7	19.7/8.8	McDonald (1977)
Kenya, village	0.83		2.7	12.3	Trpis and Hausermann (1986)
	0.89[1]		5.9	19.8	
Rio de Janeiro					Maciel-de-Freitas et al. (2007b)
suburb	0.76[2]		2.5	8.4	
favela	0.87[2]		5.0	16.5	

Note: Estimates use the Fisher-Ford model except where noted. The 50% and 10% columns show the expected number of days to survival for 50% or 10% of the population; the first number is females, the second is males.

1. Adjustments in Reiter 2007.
2. Exponential model of Buonaccorsi et al. 2003.

(2007) found a daily survival rate of 95% at two sites in coastal Kenya. Neither study, however, distinguished sexes or the taxa within *gambiae* s.l. J. Matthews et al. (2020) review a large number of studies across the genus *Anopheles* with particular attention to differences among methods used. Those done on the *gambiae* complex indicate a daily survival rate of 80% to 90%.

Culex pipiens

The most thorough study of longevity in this complex is for *pipiens* s.s. (Jones et al. 2012). For females, they estimate a daily survival rate of 90% in a suburb of Washington, DC. A lower survival rate, estimated to be 65%–84% per day, was determined for *quinquefasciatus* at a number of sites in California (Reisen et al. 1991).

Summary

A few generalities can be gleaned from these limited data. Daily survival of these three mosquitoes is between 70% and 90% in nature. Females are longer lived than males, with 90% of males dead by 7 days, whereas 90% of females are dead by 15–20 days. These numbers are much less than may be obtained in optimal laboratory conditions. For example "life expectancy" (50% mortality) of 74 days was observed for *quinquefasciatus* (Suleman and Reisen, unpublished data, cited in Jones et al. 2012; see also figure 7.1 for other comparisons of laboratory and field survival of adult mosquitoes). Survival rates are discussed further in chapter 7.

Non-Blood Feeding

While ingestion of vertebrate blood is the most distinctive feature of mosquitoes, adults also acquire energy and nutrition by ingesting other fluids, especially flower nectar or other dilute sugar solutions. This is the only source of nutrition for males. In the laboratory, *aegypti* males die within four days if not provided with sugar (Chadee et al. 2014). Nectar also supplies other resources, such as amino acids, vitamins, salts, metals, and sterols (Rivera-Pérez et al. 2017). Attraction to flowers is guided by several compounds—alcohols, aldehydes, phenols, and terpenes—some of which are produced by microbes associated with nectar (Nyasembe and Torto 2014). Lahondere et al. (2020) have identified *aegypti* neurons involved in floral attraction.

Lehane (2005) has a lengthy discussion of possible evolutionary scenarios for ancestral lineages evolving from phytophagy to haematophagy.[2] Particularly intriguing is the suggestion that the elongation of mouthparts to obtain nectar located deep within flowers was a preadaptation to using these pointed mouthparts to pierce vertebrate skin.

Mosquitoes are often accused of being nectar thieves—stealing nectar without providing anything to the plant. There are cases, however, of mosquitoes acting as pollinators for plants. Some orchids have long been known to be pollinated by mosquitoes (Dexter 1913). Peach and Gries (2020) provide a particularly convincing case that *pipiens* acts as a pollinator for several plants in the family Asteraceae. At a site in Canada, 25%–50% of field-collected *pipiens* carried pollen from at least one aster.

Peach and Gries (2020) offer a comprehensive review of mosquito phytophagy. The issue of whether any of these three mosquitoes is an essential pollinator for any plant is addressed further in chapter 7, when ecosystem function is discussed.

Notes

1. Emil Goeldi was a Swiss-born naturalist with a specialty in mosquitoes. He spent most of his career in Brazil and produced wonderful drawings of these creatures. His son, Oswaldo Goeldi—perhaps inspired by his father's drawings—was a professional printmaker of modest renown.

2. Phytophagy is the ingestion of plant-derived materials, and haematophagy is the ingestion of vertebrate blood.

CHAPTER 6

Blood Feeding

> The best blood will at some time get into a fool or a mosquito.
>
> —Benito Mussolini

Origin of Blood Feeding in Mosquitoes

Adults of most mosquito species obtain nutrition from nectar (chapter 5), and it has been suggested that elongated mouthparts for reaching nectar in narrow flowers was a preadaptation to piercing vertebrate skin for blood meals (Lehane 2005). The diversification of the major subgenera of Culicidae coincided with the diversification of land plants (Soltis et al. 2019; Soghigian et al. 2023). While nectar remains the major source of nutrition for male mosquitoes, it is not sufficient for females to develop eggs, certainly not the 100+ eggs that are formed after a blood meal. Entomophagy (sucking hemolymph from insects) may also have been a pathway to taking blood from vertebrates. Indeed, *aegypti* (and *Culex tarsalis*) have been documented to suck hemolymph from caterpillars and even obtain sufficient nutrients to produce eggs (Harris et al. 1969).[1] Taking a blood meal specifically for egg production is advantageous, however, as it allows faster larval development, because the acquisition of sufficient *nutrients* to produce eggs is not required before eclosing.[2]

Table 6.1 Summary of blood meals from field-caught females of our three mosquitoes

Taxon	No.	Humans	Non-human mammals	Reptiles	Birds
Ae. aegypti s.l.	1,615	77.6%	20.9%		1.5%
C. pipiens	13,300	37.5%	30.8%	<1%	31.7%
C. quinquefasciatus	18,233	11.0%	42.2%	<1%	57.7%
An. gambiae s.l.	2,737	39.4%	60.2%		0.4%

Note: These studies range over a number of years, with many of them done before taxonomic issues were clarified. Thus species designations are somewhat loose. Regardless, the patterns of *gambiae* and *aegypti* preferring mammals and *pipiens* (including *quinquefasciatus*) preferring about equally birds and mammals are clear. From Soghigian et al. (2023).

Female mosquitoes weigh about 2 mg, and they ingest 3–5 mg of blood when fully fed, considerably reducing their mobility. In fact, for about 24 hours after blood feeding, females remain immobile unless disturbed, resulting in increased exposure to predation. They are lighter and more mobile before a first blood meal, as well as between blood meals, since much of the additional weight is shed in fluid excretion, digestion, and oviposition.

Soghigian et al. (2023) compiled the most extensive database of records of blood meals obtained from mosquitoes captured in the field: 293,308 individual determinations from 422 species of mosquitoes (table 6.1). When the species included in these data are placed in a phylogenetic context, it becomes clear that amphibians were very likely the first vertebrates to serve as mosquito blood hosts (figure 6.1). Consistent with this are data that Corethrellidae, the closest blood-feeding insect relatives of Culicidae, feed exclusively on amphibians (Borkent 2008). About 120 MYA (mid-Cretaceous), members of the subfamily Culicinae (ancestors to *aegypti* and *pipiens*) shifted to reptile feeding. *Aedes* switched to mammals around 80–90 MYA. Thus this genus's diversification coincides with the diversification of mammals. *Culex* shifted to birds about the same time, or continued to feed on reptiles, if we accept the notion that birds are simply modified reptiles. Anophelines seem to be the first lineage to primarily use mammals, beginning about 150 MYA.

Today, with respect to our three mosquitoes, we can generalize that

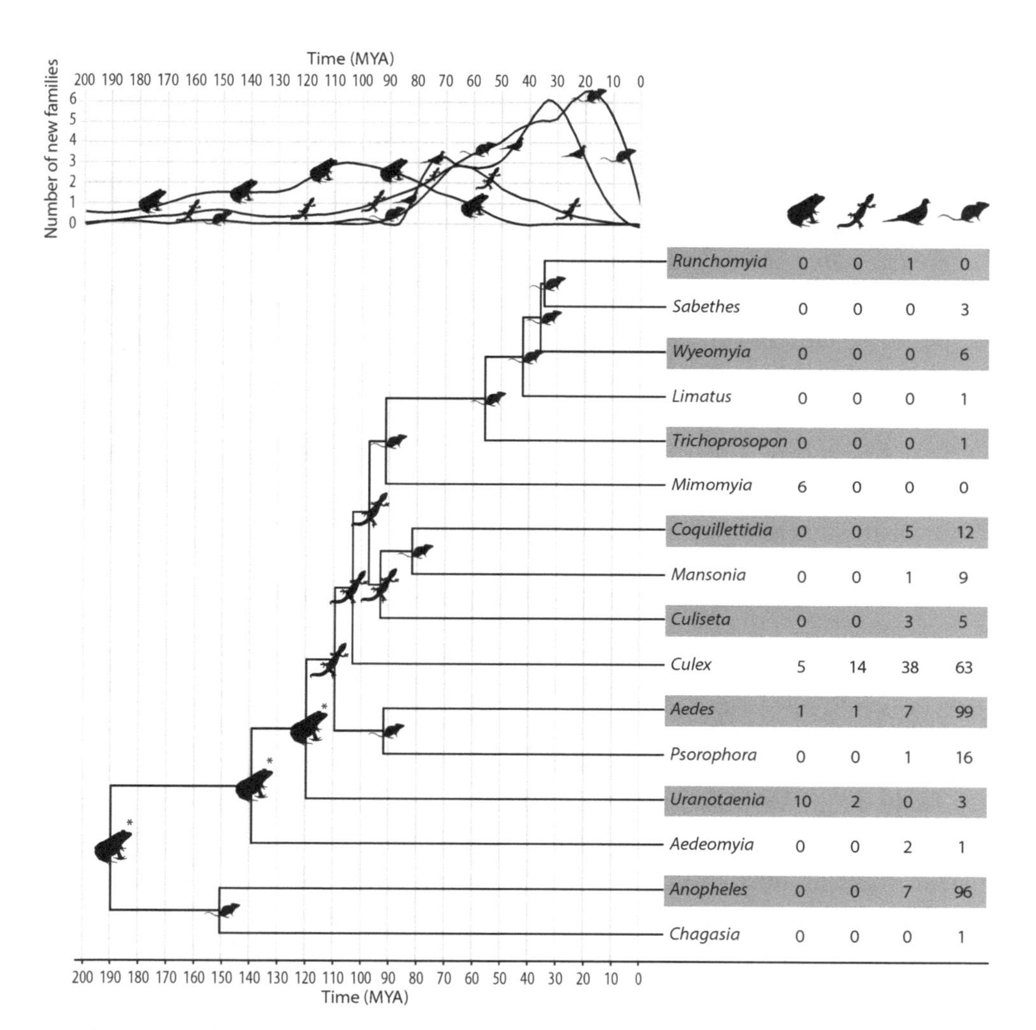

Figure 6.1 Phylogenetic tree of the major genera of mosquitoes and their predominant vertebrate host. **Upper left**: The graph shows the number of new vertebrate families appearing in the fossil record throughout the same time period when mosquito blood meal choices evolved. Note that the deepest branch involves amphibians, with anophelines branching early to prefer mammals. Most lineages remain amphibian or reptile preferring until later. While the predominant vertebrate from which blood meals are taken is shown on the branches, variations within the taxa exist. **Right**: The table at the tip of each branch shows the number of species in each group found to take blood meals from the four different types of vertebrates. Times shown at the bottom are derived from DNA coalescence estimates. This is a simplification of figure 2C in Soghigian et al. (2023), kindly redrawn by Gen Morinaga.

aegypti and *gambiae* predominantly use mammals as blood sources, while *pipiens* s.l., especially *quinquefasciatus*, use birds as well as mammals. Ribeiro (2000) has argued that the use of mammals by *quinquefasciatus* is relatively recent, as this species is much more efficient in obtaining blood from birds. The previous paragraph discussed general trends in *predominant* hosts, but it needs to be emphasized that all three mosquitoes can feed on vertebrates other than their primary victims, depending largely on availability (further noted in figure 6.1). This is especially important in zoonotic diseases like West Nile virus, which circulates among birds but can infect humans when a *pipiens* female uses first a bird and then a human for a blood meal. Such species are considered bridge vectors, moving a pathogen from a primary vertebrate host to another type of vertebrate.

Common Characteristics of Blood Feeding

The taking of vertebrate blood by mosquitoes is an elaborate process, involving complex roles for anatomy, chemistry, and behavior. Some aspects of blood feeding are common to all three of our mosquitoes. Four reviews can be consulted for more details: Besansky et al. (2004), Takken and Verhulst (2013), McBride (2016), and Coutinho-Abreu et al. (2022). Fikrig and Harrington (2021), however, critique methods for assessing host choice in mosquitoes and stress the need for caution in reaching firm conclusions.

During blood feeding, the labrum, which acts as a sheath for the delicate fascicle (needlelike mouthpart), is folded back, allowing the needle to enter into the skin and then a capillary (figure 6.2). This needle has two tubes: one leading directly to the gut, through which blood will flow, and the other coming from the salivary glands. The former is connected to a pump in the head, to facilitate blood intake. The tube from the salivary glands injects a number of chemicals (mostly proteins or peptides), including a numbing agent, a vasodilator, and an anticoagulant (reviewed in Ribeiro and Francischetti 2003; also see Ribeiro and Arca 2009).[3] These proteins are what set off an immune reaction in mammals, resulting in inflammation and itching. More importantly, most human pathogens transmitted by mosquitoes must make their

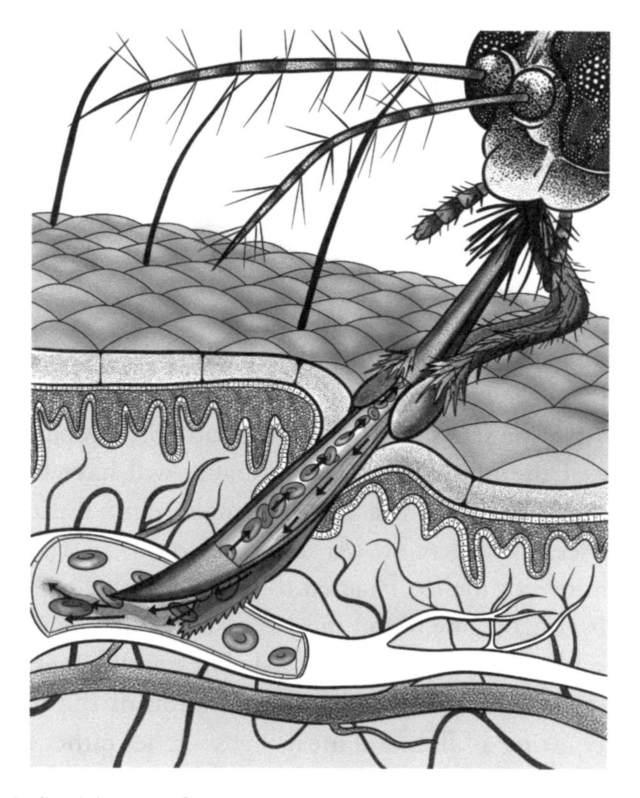

Figure 6.2 Stylized drawing of a mosquito taking a blood meal. Note that the enlarged proboscis has two tubes, one injecting fluid from the salivary glands, and one sucking blood from the capillary of a host, which eventually travels to the mosquito's gut. Drawing by Jacquelyn LaReau.

way to the salivary glands in order to be injected into a new host. This process takes several days (chapter 11).

A female needs to acquire a certain volume of blood to switch from blood-seeking behavior to resting and egg development. If disturbed, or if insufficient blood is obtained, the female will take additional blood meals before egg development can be completed. (Later in this chapter, the frequency of multiple blood meals during a single gonotrophic cycle is discussed.)[4] Blood meal size is controlled by the degree of midgut distention, mediated by the ventral nerve cord. Cutting this nerve results in females ingesting too much blood, even to the point of exploding (Gwadz 1969).

Once in the midgut, the blood bolus is surrounded by a thin membrane (the peritrophic membrane), the function of which is unknown. Much of the fluid plasma in this bolus is excreted out the rectum in the form of a small anal droplet, leaving a mass of cellular material. Degrading enzymes—especially proteases, as well as nucleases—are induced and digestion ensues, being largely completed in 24–36 hours.

Host Choice

A female's behavior changes rather drastically during the gonotrophic cycle. Initially, after eclosing, she remains quiescent for two to four days, after which she starts looking for a blood host. Depending on the local density of the population, she probably will have been inseminated after four days, with the acquired male seminal fluid also promoting host-seeking behavior. (At least for *aegypti*, if a female is not already inseminated, males attracted to blood sources are likely to accomplish this task.) If one blood meal is insufficient, she will continue her quest for blood hosts until fully fed. For anophelines, two blood meals are almost always required for maturation of the first batch of eggs. After acquiring a full blood meal, she switches rather quickly into a quiescent digesting phase for two to four days, after which she goes into oviposition-seeking mode. After laying at least most of the mature eggs, she starts seeking a blood host again and begins another gonotrophic cycle.

The particular vertebrate a mosquito chooses to victimize with its proboscis varies among and within the three species. Yet there are four common factors in attracting mosquitoes in host-seeking mode: CO_2, dark colors, lactic acid, and heat. CO_2 acts at a relatively long distance (up to at least two meters) and activates host seeking. As females come closer to a potential host, visual cues and chemicals emitted by a host and its associated skin microbiota fine-tune the choice.[5] Each taxon will be discussed separately, although it is important to note that knowledge about host choices and, especially, the underlying physiology for them is very uneven.

It is also important to distinguish between the behavior exhibited during host seeking and the actual blood meal taken. The common way to determine host-seeking behavior is by using wind tunnels in the laboratory or similar devices outdoors. Analyses of the blood in females' guts reveal the actual host(s) from which a blood meal has been taken. While these two phenomena must clearly be correlated, it is not necessarily an absolute association. Once attracted to a host, more subtle short-range cues, including using sensory neurons that rely on physical contact (touching the host), may ultimately determine whether a female feeds.

Accessibility

At least some forms of all three of these mosquitoes are closely associated with humans and can be collected both within and outside human dwellings. Figure 6.3 presents a useful summary graph of the effect of the location where females are collected on blood meals they have taken. This brings home the point that, while we can characterize these three mosquitoes as "generally" taking blood meals from a particular host, all of them are opportunistic. Their particular host choices are greatly influenced by what potential hosts are present where and when a female is hunting. In domestic settings, humans are the predominant hosts indoors, while farm animals and pets predominate outdoors. Because humans generally sleep indoors, for nighttime biters like *gambiae* s.s. and *arabiensis*, this indoor-outdoor dichotomy is especially strong (figure 6.3). Takken and Verhulst (2013) conclude, "Host preference of mosquitoes, although having a genetic basis, is characterized by high plasticity mediated by the density of host species, which by their abundance form a readily accessible source of blood."

Aedes aegypti

Field Studies

One generalization about the blood host choice of *aegypti* is that Aaf prefers nonhuman mammals, while Aaa prefers humans. Figure 6.4 summarizes the results from 10 studies that include both Aaa and ancestral Aaf in Africa (McBride 2016). It is clear that Aaa is more anthropophilic than Aaf, although the studies on field-collected pure Aaf (not in a hybrid zone) are few. Further, there is not a clear dichotomy

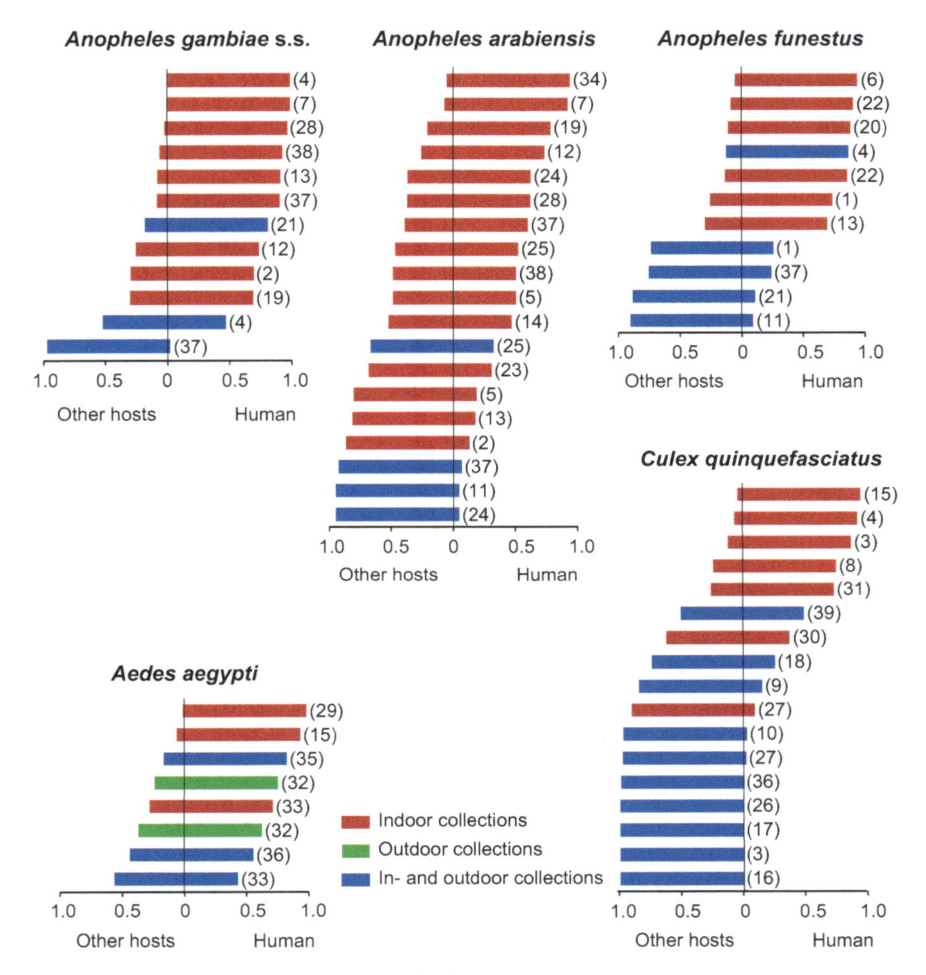

Figure 6.3 Blood meal determinations for females collected in various environments. Note that, generally, females collected indoors were more likely to take a blood meal from a human, compared with conspecific females collected outdoors. From Takken and Verhulst (2012).

between Aaa and Aaf, and host choice seems to vary geographically. For example, Hawaii's Aaa are about as zoophilic as African Aaf.

The opportunistic nature of blood meal host choice in *aegypti* is nicely illustrated by comparing the studies by Ponlawat and Harrington (2005) and Estrada-Franco et al. (2020). The former paper determined blood meal composition in over 1,000 *aegypti* females in several

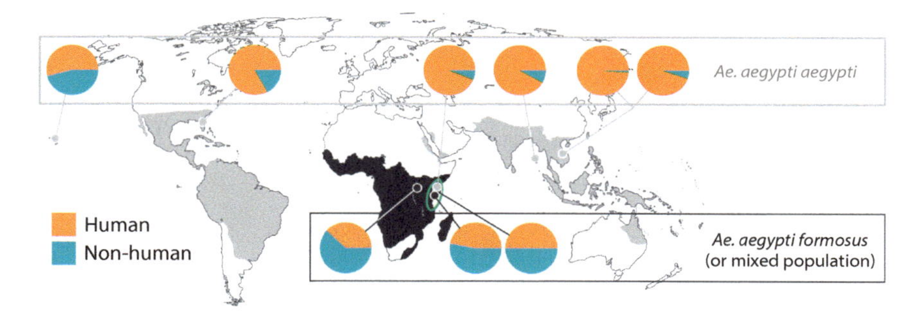

Figure 6.4 Host choices, as determined by blood meal analyses of field-captured female *Aedes aegypti*. Based on 10 studies summarized in McBride (2016). Note the paucity of studies on Aaf in Africa.

Table 6.2 Host choice for two mosquitoes common in a semirural neighborhood of a northern Mexican locale (Reynosa, Tamaulipas)

Species	No.	Single	Human	Dog	Cat	Bird	Multi	Human + Dog	Other
aegypti	145	107	30	58	18	0	38	25	13
quinquefasciatus	388	326	8	254	0	30	62	7	55

Note: Females from the field were tested for blood meal composition by PCR and DNA sequencing. Numbers of single and multiple blood meals are noted, as well as the composition of the blood for both single and multiple blood meals. From Estrada-Franco et al. (2020).

towns and cities in Thailand. About 99% fed on humans. In the latter, the authors performed their study in Reynosa in northern Mexico, in what they characterized as a semiurban, undeveloped neighborhood with poor waste management and unpaved roads. They collected *aegypti* (and *quinquefasciatus*) and used PCR primers and DNA sequencing to determine the contents of blood meals (table 6.2). Clearly, the choices for blood meals are considerably more heterogeneous in this study, compared with the data from Thailand. This almost certainly reflects the availability of vertebrate hosts. In Thailand, Aaa seems to be highly anthropophilic, and in northern Mexico, zoophilic. It is unlikely that this is due to genetic differences in the mosquitos, but instead is simply a reflection of the pool of potential blood sources available at the time and place where the studies were done.

Laboratory Studies

It is of some interest to know whether laboratory experiments on host choices of *aegypti* are consistent with what is observed in the field, to evaluate their relevance in understanding the biology of the species under natural conditions.

Generally, results from lab studies of host choices for *aegypti* are consistent with field data showing that Aaa is more anthropophilic than Aaf. McBride et al. (2014) studied collections taken in domestic settings (corresponding to Aaa) and in the forest (Aaf) in Kenya and Uganda. In laboratory studies, as predicted by their sites of capture, domestic-collected females preferred humans for blood meals, while forest-collected females preferred animals (guinea pigs, in this case). Using a combination of scaling patterns and whole genome sequencing in lab-based assessments to distinguish the two subspecies, Rose et al. (2020) found that, genetically and morphologically, Aaf females preferred animals, while Aaa females preferred humans. In chapter 7, this study is further discussed in relationship to how variations in host choices correlate with ecological factors.

Multiple versus Single Host Blood Meals

Female *aegypti* not infrequently take blood from more than one host in a single gonotrophic cycle. DNA and protein targets in immunological assays are broken down by females within about 36 hours of ingestion. Histology can also be used to differentiate vertebrate blood sources, but this method also does not work after about 36 hours (Romoser et al. 1989; Scott et al. 1993). The ability to detect multiple blood meals present at any given time is quite limited by both the technology to identify the taxa of the blood source and the amount of time between multiple blood meals. This undoubtedly results in an underestimation of multiple blood meals.

In a large study in two localities, Scott et al. (2000) determined blood meal composition from 1,891 field-caught female *aegypti* in a rural village in Thailand, and 1,675 in a residential section of San Juan, Puerto Rico. Results were reasonably consistent between sites: in Thailand and Puerto Rico, respectively, double blood meals were detected in 42% and 32% of the mosquitoes, and triple blood meals in 5% and 2%. Results were the same for females collected inside and outside houses. Along with results from northern Mexico, multiple blood meals assayed at 26% (Estrada-Franco et al. (2020), this provides strong evidence (despite technical limitations in detection) that multiple blood meals are frequent for Aaa.

Chemical Attractants and Neurobiology

CO_2 is a general, relatively long-distance stimulator for mosquito host seeking. Once this activity has begun, a number of other chemicals produced by vertebrates, as well as bacteria on skin, may affect the attractiveness of particular hosts.

Two studies are worth noting. Zhao et al. (2022) found that Aaa females have olfactory glomeruli that respond strongly to human odors but almost not at all to scents from other animals. These receptors seem specifically tuned to the long-chain aldehydes decanal and undecanal, which are enriched in human odors. The authors suggest that this sensitivity reprogramming was a crucial step in the evolution of human biting for these mosquitoes. Bellow and Carde (2022) studied combi-

Table 6.3 Compounds shown to attract or repel host-seeking *aegypti* females in laboratory experiments

Compound	Receptors	Primary effect
Carbon dioxide	GRs	activator/attractant
Lactic acid	IRs	attractant/synergist
Ammonia	IRs	attractant/synergist
Carboxylic acids	IRs	synergist
Acetone	ORs	synergist
Sulcatone	ORs	repellent
2-ketoglutaric acid + lactic acid		attractant
Decanal/undecanal	glomerulus	attractant

Note: References for the first six rows are in McBride (2016). The last two rows are from Bellow and Carde (2022) and Zhao et al. (2022), respectively. Abbreviations: GRs = glucocorticoid receptors; IRs = ionotropic receptors; ORs = olfactory receptors.

nations of compounds and found that female responses are very finely tuned to the *ratio* of various compounds. Thus mixtures of compounds are important, not simply single compounds. This sensitivity in detecting ratios of compounds extends to subnanogram dosages. (Both of these studies were conducted using the old laboratory colony of Aaa known as Orlando. It would be of considerable interest to know if recently collected strains, especially of Aaf, behave similarly.)

Vertebrate skin is well known to harbor a plethora of bacteria—a microbiome that can also produce volatile chemicals potentially detected by mosquitoes seeking blood meals. This was confirmed by H. Zhang et al. (2022), who found that the acetophenone, produced by bacteria on the skin of mice, was a strong attractant. This chemical was emitted in higher amounts in mice infected with the arboviruses causing dengue, possibly enhancing the transmission of the disease.

Timing of Blood Feeding

Ae. aegypti is generally considered a daytime feeder, although blood feeding can occur at other times. Still, the largest informative study was done by Lumsden (1957) in Tanzania. During the day, catches of mosquitoes landing on humans inside huts were 1,280, and 416 on the verandas; at night, 669 catches were indoors, and 189 on verandas. While *aegypti* have a tendency to blood feed during the day, night feeding is not infrequent.

The amount of time *aegypti* take to obtain a full blood meal has also been studied. Females, on average, spend about three minutes probing before fully inserting their proboscis, followed by taking three to five minutes to become fully engorged (Platt et al. 1997; Chadee et al. 2002).

Anopheles gambiae

Field Studies

Studies of field-collected *gambiae* s.l. indicate that this group blood feeds almost exclusively on mammals (e.g., table 6.1), although the particular mammal that is preferred varies among the taxa. As a generalization, *gambiae* s.s. and *coluzzii* feed almost exclusively on humans, while *arabiensis* is more flexible, readily feeding on domestic livestock as well as humans, depending on the availability of hosts (figure 6.3). In Madagascar, *arabiensis* feeds almost exclusively on nonhuman mammals. *An. merus* and *melas* are primarily animal feeders, although they infrequently do bite humans if humans encroach on their environment. *An. quadriannulatus* almost never feeds on humans. Little is known about *bwambae*, but it very likely is also an animal biter. (See Besansky et al. 2004 for references to the above.)

While these generalizations may often be valid, members of the *gambiae* complex are like *aegypti* in displaying opportunistic behavior. Sousa et al. (2001) studied *coluzzii* s.s. on the island of São Tomé. Of the blood-engorged females collected indoors, 93% contained human blood, while only 24% of those captured outdoors did. The predominant blood in outdoor-collected females was from dogs (52%). Osborne et al. (2018) discuss the plasticity of host choice in *gambiae* in more detail.

Given the difficulty of rearing members of the *gambiae* complex in the lab, far less detailed work has been done in identifying specific attractants. One set of studies attempting to bring the detail possible with lab-controlled work to the field, however, deserves mention. Costantini et al. (1993, 1998) devised a host-choice apparatus dubbed OBETs (odor-baited entry traps) that could measure the behavior of *gambiae* in the field. An electronic device was added later to electrocute mosquitoes attracted to each of two odor sources (Torr et al. 2008). Studies using these methods have largely confirmed the generalities discussed above. For example, in a site in Ethiopia, 46% of recently blood-fed female *arabiensis* collected outdoors had fed on humans, although cattle outnumbered humans 17:1 (Tirados et al. 2006). The OBETs caught five times as many *arabiensis* in human-baited traps versus cattle-baited ones.

Laboratory Studies

Some lab studies on the *gambiae* complex have identified the chemicals noted in table 6.3. Nonetheless, the differences observed between field-collected members of the *gambiae* complex and a widely used laboratory strain (Kisumu)—when studied under field conditions—provide reasons to question how accurately lab studies on choices of hosts for blood meals reflect the underlying biology in the *gambiae* complex (Lefèvre et al. 2009a, 2009b).

Genetics

How much of the variation in blood meals taken by *gambiae* s.l. is due to innate genetic determination and how much to variation in the environment (the pool of potential hosts) is not known. One difficulty is that many aspects of behavior in this complex are associated with chromosomal inversions, including where adults choose to spend time (e.g., indoors or outdoors), which, in turn, determines the pool of potential hosts. (The role of inversions in this group is dealt with in more detail in chapter 9.) Thus blood meals may be distributed nonrandomly among genotypes, where the underlying "choice" of host is not the determining factor. Rather, different genotypes are nonrandomly distributed among types of environments and thus encounter different pools of potential hosts.

Number of Blood Meals

Few studies have been performed on the number of blood meals found in field-collected *gambiae* in a single gonotrophic cycle. The two most informative are by Scott et al. (2006), who documented about 11% multiple blood meals in field-collected *gambiae* s.s. in Western Kenya, and by Norris et al. (2010), who observed 19% for *arabiensis* at a site in Zambia.

One characteristic of oogenesis (the formation of female gametes) in anophelines has been used to infer multiple blood meals: the frequency of pre-gravid females in a *gambiae* population. After eclosion, a single blood meal does not allow eggs to develop to maturity. Rather, oogenesis is interrupted at what is known as stage II (Gillies 1954). Females in this state are termed "pre-gravid" and will not develop their eggs to maturity without at least one additional blood meal. Briegel (1990) has attributed this to the fact that newly eclosed anopheline females have a lower nutritional reserve than culincines (*aegypti* and *pipiens*).

The pre-gravid state can be detected by dissection and examination under a microscope. Lyimo and Takken (1993) field collected observably blood-fed *gambiae* s.l. in Tanzania and held these mosquitoes for two to three days before dissecting them and examining their ovaries. The authors found that 21% remained in a pre-gravid state, while 79% became fully gravid (having mature eggs). Because multiple blood meals are required for the first batch of mature eggs, this implies that 79% had fed at least twice. The relatively low rate of multiple blood meals observed by examining the number of blood meals in field-caught females (11% and 19%, respectively), is likely due to the digestion of previous blood meals and/or that most of the field-caught samples were in their second+ gonotrophic cycle and thus did not require multiple blood meals to produce eggs.

Timing of Blood Feeding

Unlike *aegypti*, which is characterized as a daytime feeder, members of the *gambiae* complex are nighttime feeders. Bayoh et al. (2014) studied biting times (via human-landing captures) of *gambiae* s.l. and *arabiensis* at two localities in western Kenya. Overall, there is not much difference between indoor and outdoor captures, nor between species. One of the goals of the study was to see if the behavior of these species had changed, due to the introduction of insecticide-treated bed nets in these towns beginning in about 2000. There was some indication that more females were feeding outdoors in 2011 compared with 1989.

Parasite Manipulation

There have been suggestions (but little data) that females carrying a human malaria pathogen change their behavior to enhance transmission. In her review, Hurd (2003) states, "Just two studies both demonstrating that sporozoite infection cause impaired blood location and increased probing (Rossignol et al. 1986; Wekesa et al. 1992) have given rise to a perceived view in the literature that vector feeding behavior is manipulated by malaria parasites. How true is this interpretation?" (The first referenced paper is on *aegypti*.)

The confusion over this issue is illustrated by two papers appearing after Hurd's review. Nguyen et al. (2017) found no effect from a *Plas-*

modium infection on female blood feeding in *gambiae*, *coluzzii*, or *arabiensis* in laboratory studies. Later, these same researchers found no differences between infected and uninfected females among these three species in their attraction to hosts in the field, using outdoor bait traps (Vantaux et al. 2021). Blood meal analyses, however, revealed a 24% increase in human blood in females carrying the transmissible stage of *Plasmodium falciparum* (sporozoites), but not in those carrying other stages of that parasite. It is important to note that *falciparum* can only infect humans (and a few closely related primates). So, under most conditions, infected females biting animals other than humans lead the parasite into a dead end. The authors attributed the difference in the results of their two studies to the first being done in the lab and the latter in the field.

Another confounding issue in some studies claiming increased human biting by *falciparum*-infected *gambiae* females is the variation in host choices within the mosquito population. If some genotypes of the population have an innate preference for humans or a preference for habitats occupied primarily by humans (e.g., Petrarca and Beier 1992), they will have a higher frequency of human malaria infections, compared with the entire mosquito population. In subsequent blood feedings, such infected females will be found nonrandomly feeding on humans again, because that is how they became infected in the first place. The parasite has not changed (manipulated) the behavior of the mosquito.

Culex pipiens

C. pipiens is characterized as being a bird biter and, compared with the other two species complexes discussed above, *pipiens* does take blood more frequently from birds than the other mosquitoes. Farajollahi et al. (2011) provide another summary of blood meals for *pipiens* (figure 6.5), with somewhat different results than in table 6.1. (Table 6.1 is based on larger numbers, with data available up to 2022.) One important difference is that the latter set of data indicate *pipiens* s.s. takes blood meals more often from humans than does *quinquefasciatus*. This is probably due to *pipiens* s.s. being more temperate in its distribution and thus

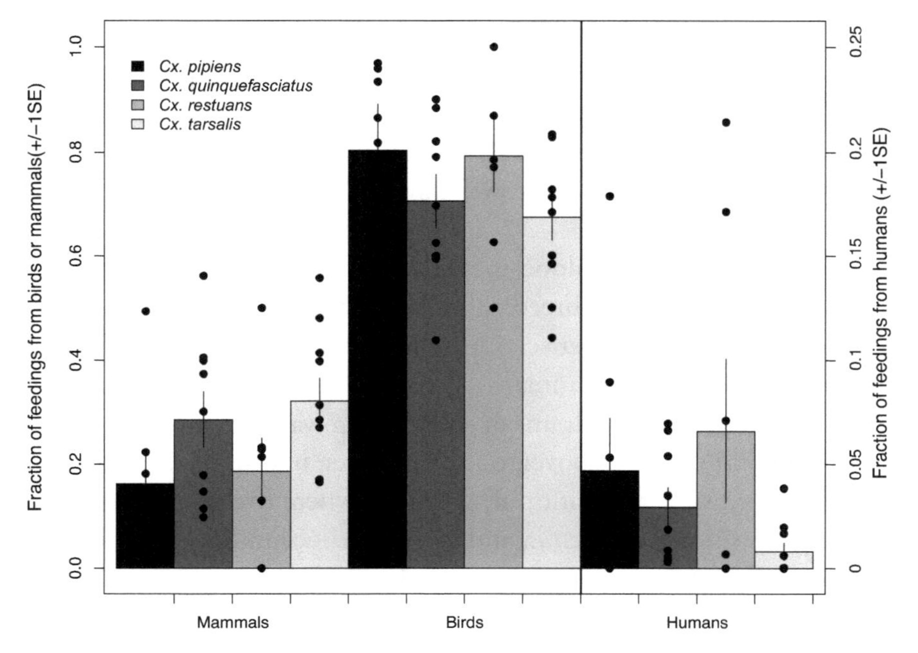

Figure 6.5 Host choices for four species of *Culex*. The first eight columns use the left y-axis scale, and the next four columns use the right y-axis scale. The four bars on the right are a subset of the four bars on the left. As an example of how to read this graph, host choices for *pipiens* s.s. (darkest shading on the left-hand set of bars) averaged about 18% mammals and 81% birds. Of the mammals, 4% were human (darkest shading on the right-hand set of bars). Note the scatter of points from studies done in different localities, which have different arrays of potential hosts. From Farajollahi et al. (2011).

likely to spend a greater amount of time indoors and be exposed to a larger number of humans than the more tropical *quinquefasciatus*. But when *quinquefasciatus* comes indoors, it is more anthropophilic (figure 6.3).

The opportunistic nature of members of the *pipiens* complex, like that of the other two mosquitoes, is indicated by the northern Mexico study that included *pipiens* as well as *aegypti* data (table 6.2). In this environment, when presented with an array of potential hosts in different abundances, *quinquefasciatus* took most of its blood meals from dogs, rather than birds or humans, as would be expected from data in table 6.1 and figure 6.5. Another indication of this flexibility is the scatter of points in figure 6.5, where each dot represents a separate study in disparate localities with different sets of potential hosts. While the above compilations of data indicated *pipiens* s.s. is less prone to take blood from birds than *quinquefasciatus*, studies in the United States (Virginia and Chicago) found that, respectively, 91% and 99% of the analyzed *pipiens* s.s. blood meals were avian (Kothera et al. 2020; Khalil et al. 2021).

Less is known about blood meal choices in *molestus*. At higher latitudes, where this form breeds almost exclusively underground and is autogenous (see the next section), *molestus* is generally considered a mammal biter, including humans (Vinogradova 2000). Presumably few birds are found in storm drains or subways. In warmer climates, *pipiens* s.s. and *molestus* coexist aboveground and often hybridize. Gomes et al. (2013) studied such a situation in Portugal, where (by genetic analysis) pure *pipiens* s.s., pure *molestus*, and hybrids co-occur. Regardless of taxonomic affinities, more than 90% of the analyzed blood meals were avian.

Number of Blood Meals

The highly urban-adapted, often subterranean-breeding *molestus* is unique among our three mosquitoes in being autogenous—being able to produce an initial batch of viable eggs without a vertebrate blood meal. This is the converse of many anophelines (including *gambiae*), which require multiple blood meals to produce an initial batch of viable eggs.

Similar to the study on blood host choices by Gomes et al. (2013), Arich et al. (2021) examined populations in northern Africa (Morocco), where anautogenous *pipiens* s.s. and autogenous *molestus* are less distinct and hybridize. In fact, several microsatellite alleles, diagnostic of each form in higher latitudes, were in Hardy-Weinberg proportions, indicating random mating. The frequency of autogeny in the hybrid populations was quite uniform (15%–18%), with no correlation with form-specific alleles.

Multiple blood meals have also been documented in *pipiens* s.l. In the northern Mexico study (Estrada-Franco et al. 2020; table 6.2), 18% of *quinquefasciatus* had multiple blood meals.

Timing of Blood Feeding

Like *gambiae*, *pipiens* appears to primarily be a nighttime biter. Host-seeking behavior begins about one hour after sunset (Anderson et al. 2007; Reddy et al. 2007). Their searching activity may either be quite short or last throughout the night. There is some evidence that the length of host-seeking times may vary seasonally.

Parasite Manipulation

Members of the *pipiens* complex transmit avian malaria. Cornet et al. (2019) found that *Plasmodium*-infected and noninfected *quinquefasciatus* females did not differ in their blood-feeding behavior, although infected mosquitoes took larger, longer blood meals, compared with uninfected ones. Conceivably, the longer the female has her labrum in a vertebrate, the better the chance for a parasite to be transmitted.

Notes

1. It has also been suggested that imbibing vertebrate eye secretions was an intermediate step to taking blood (Ribeiro and Arca 2009). There are extant gnats and moths that do this.

2. Azar et al. (2023) examined the oldest mosquito fossils known (60–100 MYA) and interpreted the mouthparts on two male fossils to be piercing—thus able to take blood meals. While intriguing, this interpretation is controversial. Especially questionable is whether the type of food imbibed by mosquitoes can be inferred by mouthparts alone.

3. These secretions from the salivary glands are not required for mosquitoes to obtain a blood meal, as females with their salivary glands removed can still ingest blood (Rossignol and Spielman 1982). This seems true, however, only if females have long periods of contact with the hosts. Timing the length of feeding indicates that these secretions significantly speed up blood acquisition (Ribeiro 1988).

4. The cycle of blood feeding → egg maturation → oviposition → blood feeding is called the "gonotrophic cycle." Long-lived females may undergo several such cycles in a lifetime.

5. Beer consumption has also been implicated as increasing attractiveness, a seemingly amusing point, except that alcohol consumption is accelerating in many tropical regions. The increased frequency of blood feeding this induces can have epidemiological consequences (Lefèvre et al. 2010).

CHAPTER 7

Ecology

> The first law of ecology is that everything is related to everything else.
>
> —Barry Commoner

Several aspects of the biology of these three mosquitoes that might fall under a broad definition of ecology have been discussed in previous chapters and will not be repeated here: biogeography (chapter 1), larval feeding and breeding sites (chapter 4), and adult feeding and active dispersal (chapter 5). In this chapter, other aspects of ecology will be discussed, albeit not in depth. Three books—Service (1976), Lounibos et al. (1985), and Charlwood (2020)—totaling more than 1,500 pages, go into much more detail. Despite this extensive literature, it is important to note that many aspects of the ecology of mosquitoes cannot be understood—much less mastered—solely by reading ("book learning"). Anyone seriously interested in pursuing mosquito ecology should go into the field with an experienced field biologist who is working with the species of interest.

Collection Methods

A first step in studying the ecology of mosquitoes in the field (where ecology is to be found) is to find them. Various methods have been used to attract and collect our three mosquito species, each with idiosyncrasies that are worth noting. All stages of these mosquitoes (eggs, larvae/pupae, and adults) can be collected with varying degrees of ease and success.

Eggs

For *aegypti*, making egg collections is generally the preferred method, especially if collections are meant to be shipped or brought back to a remote lab. Ovitraps (egg traps) consist of partially water-filled containers equipped with a wooden paddle or lined with sturdy paper (seed germination paper used in plant nurseries works well). Flat-black containers work best. Placing several of these both inside and outside buildings, tire dumps, and/or cemeteries for at least two days should attract ovipositing *aegypti* females, if they are present. The longer a trap is in the field, the more natural microbiota it will accumulate, adding to its attractiveness. This can be speeded up if naturally acquired standing water can be added to the trap.

One issue with relying on one or a few ovitraps is the risk of biased sampling, due to the eggs being siblings—that is, the eggs in one trap having all been laid by one or very few females. Female *aegypti* are skip ovipositors (chapter 3), laying a few eggs in one location and moving on to another, minimizing the number of siblings at a single site. Nevertheless, if collections are to be used for studies that assume random sampling, being cognizant of this issue is important. Collections from multiple ovitraps, spaced at a certain distance (100+ m), can minimize sibling samples, especially if only one or a few eggs are sampled from each trap. I have gone into some detail on the nature of *aegypti* ovitraps because, in addition to being employed to obtain samples in the field, they are also used to obtain a crude measure of relative population sizes for, among other purposes, monitoring the effectiveness of control programs (see the next section).[1]

Eggs rafts for *pipiens* are relatively noticeable and can be easily collected. Only *Culex* and the rare genus *Culiseta* lay egg rafts, so a collector can be confident of getting the right genus by using this method. *An. gambiae* eggs are more difficult to identify and collect, especially because multiple species are probably ovipositing at the same site. Neither *pipiens* nor *gambiae* eggs remain viable for an extended time, so transporting them to the lab needs to be done quickly. Service (1976) has a chapter detailing methods that have been used (with varying degrees of success).

Larvae

Methods for collecting larvae of all three mosquitoes are quite similar. The first (and often hardest) step is to locate actively breeding larval sites. Once found, if they are sufficiently large and accessible, a dipper with a screened bottom can be used. The openings of the sieve can be adjusted to collect various-sized larvae—that is, instars at different stages. If a water pool is inaccessible or small, a baster can be used to suck the water into a clear jar and then be examined for the presence of larvae. Or a tube can be used to set up a siphon for situations like a tree hole high in a tree. Service (1976) also has a full chapter (77 pages) devoted to collecting larvae. Determining the species of larvae can be difficult, as taxonomy is most often studied in adults. Larval keys do exist, but it may be more convenient to rear immatures to adults or use DNA sequencing (www.barcodinglife.org).

Adults

Adult mosquitoes are usually collected by one of three categories of methods. The first is human-landing catches (HLC). A collector exposes skin and simply picks off adults as they land to blood feed. While the majority of captures are females (presumably seeking a blood host), a minority of males is often collected. This is especially true for *aegypti* (chapter 3), as it is likely to be a way males find females for mating. The ethics of this method can be complicated, especially when performed in a locality where there is active transmission of diseases that are vectored by the mosquitoes being collected.[2]

Double-net contraptions have also been used, where a human or an animal is not directly exposed to mosquitoes, but simply used to attract them into a net. Often a current of air is blown across the animal and funneled into a trap, such as OBET traps used to study attractants (chapter 6).

For members of the *gambiae* complex that feed indoors at night, recently blood-fed females can be collected by aspirating resting females on the interior of huts early in the morning. Females taking blood meals generally fly only as far as a wall to begin digestion. If polytene chromosome analyses are the goal of the study, this is an especially convenient way to obtain a sample of gravid females with nurse cells appropriate for chromosome preparations.

Several adult traps that do not use live animals (or humans) have also been devised. CO_2 is an attractant that acts at a relatively long range (chapter 6). It may be generated in traps by using gas tanks, dry ice, or fermenting yeast. Light—both visible and UV—may attract some adults. Various other attractants, such as hay infusions and concoctions meant to mimic human odors, have also been developed.

A number of studies have been performed trying to evaluate the effectiveness of the multitude of adult mosquito traps. A partial list of relatively recent studies includes Krockel et al. (2006), Y. Li et al. (2016), Kenea et al. (2017), Mburu et al. (2019), Mwanga et al. (2019), Abong'o et al. (2021), van de Straat et al. (2021), Goi et al. (2022), and Zembere et al. (2022). To summarize, no single trap is most effective in all environments for all species of mosquitoes.

Before leaving the discussion of collecting mosquitoes, some advice: if at all possible, make contact with local mosquito workers in the area where you intend to collect. They will know where and when to find the species you are seeking. In addition to academic researchers, many regions have government mosquito control agencies with experienced personnel who know the local conditions well.

Population Sizes

Methods

Population sizes can be measured either as relative size or absolute size. Relative sizes of mosquito populations are estimated by various indices: the fraction of traps registering positive for a species, the number captured per trap, the fraction of examined water pools that contain larvae, the number of successful ovitraps, and so forth. Tun-Lin et al. (1996) and Focks (2003) discuss a number of such relative measures. Williams et al. (2008) introduced a variant where sentinel breeding containers are purposely deployed and provide simulation models to calibrate the results.

For some purposes, estimating the relative population sizes of mosquitoes is adequate. This is especially true for evaluating the effectiveness of various control measures aimed at reducing a population. To be convincing, such estimates need to be performed over a significant length of time—both before and after a treatment—to document whether the control program has truly reduced populations, or an apparent reduction is just a natural fluctuation, independent of any control measures.

For estimating absolute sizes, as well as survival rates (see the next section), animal ecologists, including insect ecologists, have long considered mark-release-recapture (MRR) as the standard (Southwood 1978; Southwood and Henderson 2000). As the name implies, this involves marking a known number of mosquitoes of the species of interest, releasing them into the field population, and then recapturing them. In its simplest conception, if the released individuals randomly mix with the field population, and the recapture technique equally attracts marked and wild mosquitoes, then the ratio of marked to unmarked individuals in the recaptured sample should be equal to the ratio of the number of released mosquitoes to the size of the field population. Many details are involved, including the various analytical methods used, and complications can arise.

One of first issues is the source of the released individuals. Laboratory-reared mosquitoes have been used, sometimes marked with a distinguishable morphological mutation. Using a sample directly from the field population, however, is preferable, to avoid laboratory-rearing artifacts.

The second issue is how to mark the mosquitoes. Various methods have been used, with differing degrees of sophistication and difficulty, such as feeding mosquitoes with dyes or radioactive isotopes (Service 1976). One very important assumption is that the way in which the mosquitoes are marked does not negatively affect their behavior (e.g., causing increased death or reducing capture rates).

Fortunately, there is a fairly easy way to mark mosquitoes, using UV-fluorescent dusts that come in various colors. A cage of mosquitoes captured in the field can be puffed with the dust, producing a cloud, or the inside of cups used to hold the mosquitoes can be brushed with the dust. This has the advantage of minimally handling wild mosquitoes, as well as allowing researchers to use more than one color to distinguish different days or locations of releases. Recaptured samples are then viewed under UV light to reveal the marked individuals. In a study on *gambiae* in Mali by Touré et al. (1998b), using the fluorescent dust method, recapture rates were between 4% and 10%, which were higher than those for previous studies using different marking methods, thus confirming that this method minimally (if at all) affected survival or behavior. Dickens and Brant (2014) more thoroughly studied the effects of fluorescent dusts on mosquito survival.

Analyses of the results from MRR studies may be reasonably simple and straightforward when using classical methods, such as the Lincoln Index or Fisher-Ford method. The former estimates the size of the population as N = (Mn)/m, where M is the number of marked mosquitoes released, n is the total number of mosquitoes recaptured, and m is the number of recaptured marked individuals. Fisher (1947) modified this to account for mortality between release and recapture. Fisher-Ford estimates are consistently larger than those using the Lincoln Index. Chapter 14 in Silver and Silver (2008) goes into much more detail on the technical issues (e.g., methods of marking) and analyses (e.g., stochastic vs. deterministic models, the effect of emigration, etc.).

Results

Aedes aegypti

Guerra et al. (2014) list 163 MRR "records" for *aegypti*, 87 of which had disaggregated data.[3] These authors also calculated the mean recapture rates for these MRR studies to be 11%. Notably, the recapture rate was higher in studies performed nearer the equator, compared with cooler

Table 7.1 Estimates of adult *aegypti* population sizes from MRR studies

Locality	Method of marking	Population size estimate	Survivorship	Reference
Kenya, village	FP	650 per 34 houses	0.77/0.89[1]	McDonald (1977)
Kenya, village	FP	337–464		Trpis et al. (1995)
Kenya, village	FP	635–1200		Lounibos (2003)
Cairns, Australia	*Wolbachia*	5–10 per house	0.7/0.9	Ritchie et al. (2013)
Rio de Janeiro	*Wolbachia*	3,000–6,000	0.82/0.89	Garcia et al. (2016)
New Delhi	FP	9,000–15,000		Reuben et al. (1973)
Tanzania	paint	250 females		Conway et al. (1974)

Note: Survivorship estimates are for females, except where noted. Abbreviation: FP = fluorescent powder.
1. Males/females.

climes at higher latitudes. Given this large number of studies, only a few illustrative examples are in table 7.1.

By far the most thorough monitoring of population size for *aegypti* populations was carried out by Sheppard et al. (1969) in Bangkok. They performed 23 releases over a full year, with about two releases per month. The estimated male population size varied from 350 to 2,800, and for females, from 362 to 2,200. This emphasizes that even in tropical climes with relatively stable year-round temperatures, *aegypti* populations can fluctuate dramatically, almost certainly due to rainfall variation.

Most striking in table 7.1 is the indication that adult population sizes of *aegypti* are much smaller in villages in Africa, compared with large cities outside Africa. It is not possible to attribute these disparities solely to differences between Aaa and Aaf. The populations in Kenya are reintroduced Aaa (chapter 9). The one study likely to be all Aaf is for Tanzania, although neither morphological nor genetic data were reported confirming that the population studied was Aaf.

In addition to these estimates of *aegypti* adult population sizes, attempts to estimate immature population sizes have been made with varying degrees of success (Focks et al. 2000; Morrison et al. 2008; Williams et al. 2008). It is unclear how these estimates relate to adult population sizes.

Table 7.2 Estimates of adult *gambiae* population sizes from MRR studies

Locality	Species	Release population	Population size estimate	Reference
Burkina Faso	*coluzzii*	FP lab-reared and field-caught	dry: 10,000–50,000[1] wet: 100,000–500,000[1]	Epopa et al. (2017)
Sudan	*arabiensis*	FP lab-reared	32,500	Ageep et al. (2014)
Mali	*gambiae* s.s.	FP field-caught	dry: 1,000–5,000 wet: 5,000–200,000	Baber et al. (2010)
Mali	*gambiae* s.l.	FP field-caught	year 1993: 9,000–11,000 year 1994: 28,000	Touré et al. (1998b)
Burkina Faso	*gambiae* s.l.	FP field-caught	150,000–300,000	Costantini et al. (1996)

Note: Dry and wet refer to seasons. Abbreviation: FP = fluorescent powder.
1. Males only.

Anopheles gambiae

Several MRR studies have been performed on the *gambiae* complex. Many were done to estimate survival, but some include estimates of adult population sizes, the results of which are in table 7.2. Two points from these studies are worth emphasizing. First, estimates from studies done during the dry season are about 10% of wet season estimates in the same locality. Second, these estimates are at least an order of magnitude larger than those obtained for *aegypti* (table 7.1).

Culex pipiens

Few MRR studies aimed at estimating population size have been done on the *pipiens* complex. MacDonald et al. (1968) estimated a population size of about 1,000 for *quinquefasciatus* (called *pipiens fatigans* in that paper) in Yangon, Myanmar (Rangoon, Burma, at that time). Laporta and Sallum (2008) estimated about 29,000 adult *quinquefasciatus* in a park in São Paulo, Brazil.

Extreme year-to-year variation in *pipiens* s.s. population sizes was documented in northern Italy by Rosa et al. (2014). They used CO_2 traps across a large area (about 1,000 square km) over 11 years. The number of adults captured per trap varied between about 40 and 4,000 across the years. Early-season temperatures were implicated as the driving factor in this extreme variation (Mariani et al. 2016). Reisen et al. (1991) estimated a population density of *quinquefasciatus* females as 36,000–672,000 per square km, but it is not clear how to relate this density to population size.

Note: Here absolute census sizes have been discussed, sometimes abbreviated N_c to distinguish them from effective population sizes (abbreviated N_e). N_e is a theoretical parameter in population genetics that represents the sampling size of genes from one generation to the next. N_e estimates for these three mosquitoes will be presented in chapter 9.

Survival Rates

MRR studies can also provide estimates of daily survival rates if recaptures are done over several days. The proportion of marked individuals will decrease over time, depending on survival, allowing estimates of the probability of individual daily survival or, conversely, mortality rates. A variety of methods to calculate survival/morality rates from such data exist. Simple linear models assume constant mortality over time as well, as no removal from the populations. Nonlinear models incorporate changes over time, as well as the removal of recaptured individuals (e.g., Buonaccorsi et al. 2003). Similar to population sizes, there is no single survival rate for any of our three mosquitoes that would hold true across ecological settings. The following discussion emphasizes work on natural populations. A large number of lab studies have been done, but it is not known how lab results apply in nature. Adult female survival rates (longevity) play a central role in disease transmission (chapter 11).

Aedes aegypti

A study by Brady et al. (2013) is a convenient entrée into the literature on survival rates for *aegypti* in both laboratory and field studies. They review 141 laboratory and 50 field studies, and two generalities can be gleaned from it: females survive longer than males, and survival in the laboratory is greater than in the field. In laboratory studies, daily survivorship varied from 92% to 95% (mean = 94%) while in the field, survivorship was from 40% to 97% (mean = 71%). These authors were par-

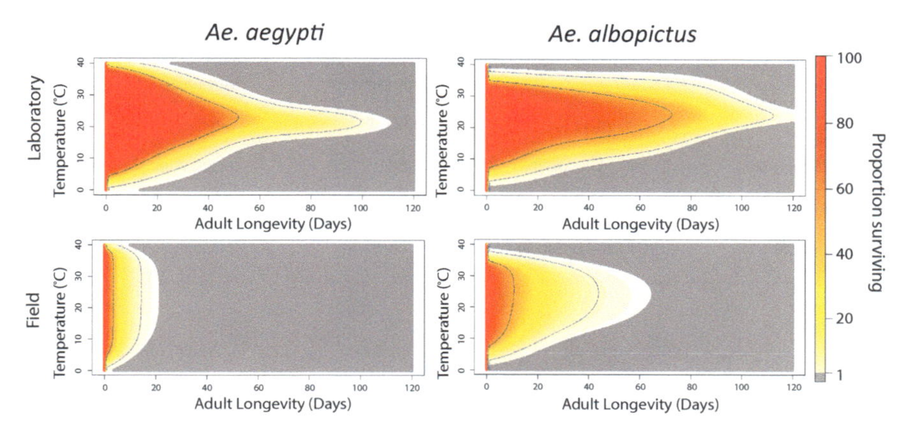

Figure 7.1 Distribution of adult female longevity at different temperatures for *Aedes aegypti* and *albopictus* from both laboratory studies and MRR experiments in the field. The shading gradient from red to white represents 100% to 1%. Dashed lines are limits for 95% and 50% of the population. *Ae. albopictus* is discussed in chapter 12. From Brady et al. (2013).

ticularly interested in the effects of temperature on survival. Figure 7.1 summarizes their work. The longer survival period measured in the lab versus in the field is very evident, as is the lowering of survival at extreme temperatures.

Harrington et al. (2008) demonstrated that mosquitoes "age" (i.e., their mortality increases with age). Mosquitoes released when they were 1–4 days old had a greater recapture rate than those 5–20 days old at release, implying that younger mosquitoes survived better than older ones. Level of nutrition also affects rates of survival in *aegypti* (Costero et al. 1999),

Anopheles gambiae

Survivorship of *gambiae* under field conditions has been estimated at three locations: in Mali (Touré et al. 1998b), Burkina Faso (Costantini et al. 1996), and coastal Kenya (Midega et al. 2007). At the two West African localities, daily survival rates of around 80%–88% were estimated, while at two localities in Kenya, daily survival was estimated at 95%.[4]

Table 7.3 Estimates of daily survival rates for *pipiens* in natural populations

Species	Locality	Daily survival	Reference
pipiens	Washington, DC	0.90	Jones et al. (2012)
quinquefasciatus	Orange, California	0.84	Reisen et al. (1992)
quinquefasciatus	São Paulo, Brazil	0.60–0.68	Laporta and Sallum (2008)
quinquefasciatus	Southern California	0.65–0.84	Reisen et al. (1991)

Culex pipiens

Few studies have been done to estimate survival rates in field populations in this group. Table 7.3 summarizes results from four studies.

MRR Summaries

To sum up results from MRR studies aimed at estimating both adult population sizes and survival rates, there is considerable variation both within and between species. This is not surprising, since these parameters depend greatly on local environmental conditions that vary both spatially and temporally. While recognizing that such fluctuations exist, we can nevertheless *tentatively* posit that, in general, *aegypti* have the smallest populations (sometimes in the hundreds), *gambiae* the largest (usually in the hundreds of thousands), with *pipiens* being intermediate (a thousand to tens of thousands). Estimates of daily survival rates are even more difficult to generalize within or between species. Depending on the study, types of analyses, season, and so forth, this parameter varies from about 70% to 95%. This means that the median adult life span can vary from 2 to 14 days, with 10% survival between 7 and 45 days, assuming a constant survival rate (i.e., no aging). When the sexes are measured separately, females survive longer than males, likely related to their larger size, implying greater metabolic reserves.

In addition to the empirical studies reviewed above, population growth and fluctuations of mosquitoes have been modeled, sometimes incorporating a large number of parameters (e.g., Otero et al. 2006).

Age Structure

While the above yields insights into total adult population sizes, it is also of some interest from an epidemiological perspective to know the age structure—that is, the proportion of adults at different ages. Estimating the age of adult mosquitoes caught in the field has been attempted in several ways. These include an examination of female ovaries to estimate the number of gonotrophic cycles, changes in cuticular hydrocarbons, protein profiling, and the like. Johnson et al. (2020) critically reviewed these methods and concluded that, "despite decades of research, defining the age of a wild-caught mosquito remains a challenging, impractical, and unreliable process." Since that review, two other methods have been proposed: quantifying wing wear (Gray et al. 2022) and quantifying the spermatozoa remaining in spermathecae (Madan et al. 2022).

Invasiveness and Anthropogenic Passive Transport

Active dispersal by adults was discussed in chapter 5. Here passive transport is examined.

Aedes aegypti

Passive transport by humans is the predominant way *aegypti* disperses over longer distances. The role of the transatlantic slave trade was noted in reference to *aegypti*'s escape from Africa (chapter 1). Two other genetics-based studies on the role of ships and boats are worth noting. A study of shipping patterns among 15 seaports on seven Philippine islands revealed that the level of genetic differentiation of *aegypti* populations at the ports was inversely proportional to the amount of interisland shipping (Fonzi et al. 2015). Of note, cargo ships were implicated as being more important than passenger ships in promoting *aegypti* migration. Guagliardo et al. (2019) performed a study of genetic differentiation along a river in the Peruvian Amazon. The level of boat traffic among communities was shown to be more important than simply the geographic distance in *aegypti* migration that is implied by genetic differentiation. Shipping is also implicated in outbreaks of temporary yellow fever along the East Coast of the United States during the 18th and early 19th centuries. Outbreaks were concentrated at the three ports receiving the most cotton: Philadelphia, New Haven, and Boston. In 1878, Memphis, Tennessee, had a particularly severe outbreak of yellow fever, almost certainly due to the introduction of *aegypti* from New Orleans on boats traveling up the Mississippi River.[5]

While the above examples confirm the role of watercraft in *aegypti* movements, at least in the Caribbean, boats generally seem to play a minor role. Chadee (1984) inspected 46,895 vessels arriving in Trinidad over a 10-year period, and only 31 contained *aegypti*.

Curiously, there is little evidence that *aegypti* has been transported by air. When introductions by air have been implicated, they do not establish permanent breeding populations (Lounibos 2002). Introductions due to land transportation (autos, trucks, and trains) are likely involved in temporary introductions of *aegypti* in northern regions of the United States in the summer (Gloria-Soria et al. 2022a), as well as permanent introductions into California (Gloria-Soria et al. 2014; Pless et al. 2017) and Washington, DC (Gloria-Soria et al. 2018b).

Anopheles gambiae

While largely remaining in sub-Saharan Africa, members of the *gambiae* complex have been passively transported at least twice, resulting in temporary breeding populations outside its native range. In 1930, *arabiensis* established breeding populations in the state of Natal in Brazil.[6] Until its eradication in 1940, this resulted in an estimated 300,000 cases of malaria, with 16,000 deaths (Soper and Wilson 1943). This introduction is thought to have been due to either a ship or airplane coming from Dakar, Senegal. Similarly, *arabiensis* was introduced into Egypt in the early 1940s, likely by airplanes from the Sudan; it was eradicated in 1945 (Shousha 1948).

Passive transport of *gambiae* by airplane is indicated by cases of "airport malaria," which occur mostly in Europe. Airport malaria is the name given to malaria cases (most often *falciparum* malaria, found in Africa) in humans living near international airports who have not traveled outside their local area. Isaacson (1989) reviewed 29 cases in Europe between 1969 and 1988, all of which were near or at airports with frequent flights from Africa. When flights are examined in detail, 8–20 anophelines (some *gambiae*) were found per flight arriving in Charles de Gaulle Airport in Paris (Gratz et al. 2000; Alenou and Etang 2021).

Wind-borne adults have also been detected (Huestis et al. 2019; Yaro et al. 2022). Night flights at altitudes between 60 and 300 meters with airplanes outfitted with insect nets captured 235 anophelines over 617 nocturnal flights, including *gambiae* s.s. and *coluzzii*. This was in the Sahel (the lower border of the Sahara Dessert) in Mali, which spans the transition from tropical forest to the Sahara Desert. In this region, surface water for *gambiae* breeding dries up for up to nine months of the year, only to have these mosquitoes reappear very quickly when rain arrives. Such reestablishment may be from these long-distance airborne adults or, as discussed later in this chapter, *gambiae* s.l. also survives extensive dry periods through subterranean aestivation.

Have there been any cases of passive transport by humans of *gambiae* that have established permanent breeding populations? The most likely instances are islands off the coast of East Africa, in particular Mauritius, Réunion, and Mayotte. These islands are thought not to have had *gambiae* present before about 1860, but today at least *arabiensis* can be easily collected, and up to three species—*arabiensis*, *gambiae* s.s., and *merus*—have been collected on some islands (Patterson 1964; Bryant and Gebert 1976; Iyaloo et al. 2014; G. Le Goff, personal communication).

Culex pipiens

Given the history of this complex (chapter 1), it is clear that several intercontinental expansions occurred at a time when boats were the only human-associated mode of oceanic transport. More-recent passive transport of *pipiens* has largely come from studies of the spread of genes associated with insecticide resistance. These genes (resistant alleles) are thought to have arisen only once (i.e., are monophyletic), so the presence of such alleles in multiple populations must be due to migration. Resistant allele A2-B2 was first detected at or near airports in California (Raymond et al. 1987), Marseille (Rivet et al. 1993), and Barcelona (Eritja and Chevillon 1999). Raymond et al. (1991) show that this spread has been worldwide. The association of initial introductions with airports implicates air travel, which was directly documented by Highton and van Someren (1970).

Competition

Competition between mosquito species most likely occurs at the larval stage, and most of the direct studies of mosquito competition have been performed in the laboratory. Here I will confine the discussion to what is known in field populations. The existence of competition in nature can be inferred under some conditions—for example, when an invasive mosquito spreads to a new region and a resident species disappears or becomes greatly reduced.

Subspecies Competition

A particularly intriguing situation in the Rabai District in eastern Kenya has allowed some detailed understanding of how the subspecies of *aegypti* (Aaf and Aaa) can coexist with minimal competition. A dark form of *aegypti*, corresponding to Aaf, was found breeding in tree holes in forested areas. A light form, corresponding to Aaa, was found inside huts less than 100 meters from the Aaf population, breeding in large clay water containers. That the two types are found close enough to be considered sympatric was confirmed by collecting adults inside huts in the dry season that corresponded to Aaf. This had been a stable situation, at least since the 1950s (Mattingly 1957). Population genetics analyses of collections made in 1975/6 and 2009/11 of mosquitoes taken from within huts and in adjacent forests found them to be genetically distinct, corresponding to the Aaa and Aaf groups (Tabachnick et al. 1979; Gloria-Soria et al. 2016a). These populations were the subject of a groundbreaking study of olfaction related to host choices (McBride et al. 2014). Since 2011, a centralized water distribution system has been introduced into the villages of Rabai, and water is no longer stored in large quantities indoors. Attempts to collect the light form of *aegypti* in Rabai huts in 2017 were unsuccessful. Indoor and outdoor *aegypti* are now morphologically and genetically homogeneous Aaf, with some signs of Aaa admixture (Xia et al. 2020). These studies nicely illustrate how closely tied the two subspecies are to different ecological niches. By removing one niche (water stored in huts), the two could not coexist and instead merged.

Aedes albopictus and *aegypti*

The review by Juliano (2000) is an excellent introduction to interspecific competition in mosquitoes. Although 23 pairs of species are listed as having been studied for competition, only five refer to interactions in nature, and all these involve the invasive *albopictus* species.

Ae. albopictus is a native Asian mosquito with very similar properties to *aegypti*. It has spread around the world since the mid-20th century, with its dispersal accelerating in the 1980s (C. Moore 1999; Kraemer et al. 2019; also see chapter 12). Because of its ecological similarity to *aegypti*, they compete in places where they meet. This has been most thoroughly studied in the southeastern United States, especially Florida.

In field experiments in Florida, Juliano (1998) found that *albopictus* outcompeted *aegypti* in water-filled unmounted tires. The results were dependent on density and per capita resources (leaf litter), with *aegypti* surviving only in places with high resource availability. Juliano also tested whether the outcome of these experiments was due to a larval parasite (*Ascogregarina*) and concluded that it was not. Similar experiments in Brazil came to the same conclusion (Braks et al. 2004).

One important difference between *aegypti* and *albopictus* is that the former is more urban adapted, while *albopictus* prefers greener areas, such as parks and suburbs. In some cases this has led to a differential displacement along a gradient from urban to rural, as was documented in Palm Beach County in Florida over a distance of about 15 km (Reiskind and Lounibos 2013; Hopperstad et al. 2021). Surprisingly, Juliano et al. (2004) found no evidence that the relative competitive ability of these two species differed between sites where the species coexisted and where *aegypti* had been excluded. That is, the outcomes of competition between these species are not due to genetic differences among populations of mosquitoes; rather, they depend on the environments where they compete.

A unique form of competition, leading to species displacement, involves mating and asymmetric reproductive interference. This has been hypothesized to occur where *albopictus* and *aegypti* overlap. Males pass a chemical (matrone) in their semen to females, which renders these females refractory to further mating (chapter 3). Interspecific matings between *aegypti* and *albopictus* do occur, but no viable offspring are produced. Of note, when *albopictus* is the male, the mated *aegypti* female becomes refractory to any mating with a conspecific, while an *albopictus* female mated to an *aegypti* male will subsequently mate with a conspecific male and produce offspring. This asymmetry in response to interspecific matrone would lead to an advantage for *albopictus* over *aegypti* in places where their distributions overlap (Bargielowski and Lounibos 2016; Lounibos and Juliano 2018). Such hybrids have been observed in nature at frequencies of 1%–4% (Bargielowski et al. 2015). Because hybridization between these species reduces fitness, increased mating discrimination is expected to evolve rapidly, as has been observed by Bargielowski and Lounibos (2014).[7]

Anopheles gambiae

Diabate et al. (2005, 2008) performed studies of competition between *gambiae* s.s. and *coluzzii* in Burkina Faso (called S and M forms in those publications). *An. gambiae* s.s. dominates in temporary pools generally surrounding more permanent bodies of water, while *coluzzii* is adapted to man-made, more permanent water, such as irrigated rice fields. As expected, the two species outcompeted each other in their native habitats. In the presence of predators (carnivorous insect larvae of the families Notonectidae and Dytiscidae), however, *coluzzii* outcompeted *gambiae* s.s. in both habitats.

While both *merus* and *melas* are largely confined to the two coasts of Africa, breeding in salt water, both can develop through to adults in fresh water. In parts of East Africa, *merus* is found much farther inland (figure 1.9), breeding in fresh water (Coetzee et al. 2000; B. White et al., 2011). These are regions where *gambiae* s.s. is absent or rare, suggesting that competition with *gambiae* s.s. is at least partially responsible for keeping these species clustered on the coasts.

Culex pipiens

Like *aegypti*, resident *pipiens* populations are negatively affected by introduced *albopictus*. At a site in northern Italy, Carrieri et al. (2003) examined almost 1,200 artificial containers and noted the presence, absence, or coexistence of the two species. Smaller containers (<5 liters) had a preponderance of *albopictus*, while *pipiens* dominated in larger containers (>100 liters). Both were found in medium-sized containers (10–50 liters). Leisnham et al. (2021) also present evidence that the competitive superiority of *albopictus* over *pipiens* is dependent on the type of container.

In a locality close to that used by Carrieri et al. (2003), Marini et al. (2017) studied the temporal shift in abundance of the two species. *C. pipiens* tended to peak in population size earlier in the breeding season, after which *albopictus* began to dominate. Evidently *pipiens* eggs, laid directly on water surfaces by females breaking diapause, hatch earlier than the eggs of *albopictus*, laid on hard surfaces just above water level, where water needs to rise to flood them. Once sufficient numbers of *albopictus* develop, they outcompete *pipiens*, reducing their populations later in the breeding season.

Niche Modeling and Climate Change

Niche modeling is an attempt to identify ecological factors that vary geographically (EGVs, or ecogeographic variables) that are associated with the relative abundance of different species. This is often done in attempts to explain how ecologically similar species can coexist and what limits their distributions. As many ecological factors as practical are examined: topography (e.g., altitude, water), climate (e.g. rainfall, temperature), habitat (e.g., land cover), and anthropogenic variables (e.g., cities, villages, roads). All these data are mapped using GIS (geographic information system) coordinates provided by satellite imaging. The relative and absolute abundances (if available) of different taxa are overlaid, and statistical analyses are performed to define the niche of each taxon and niche overlap between taxa.

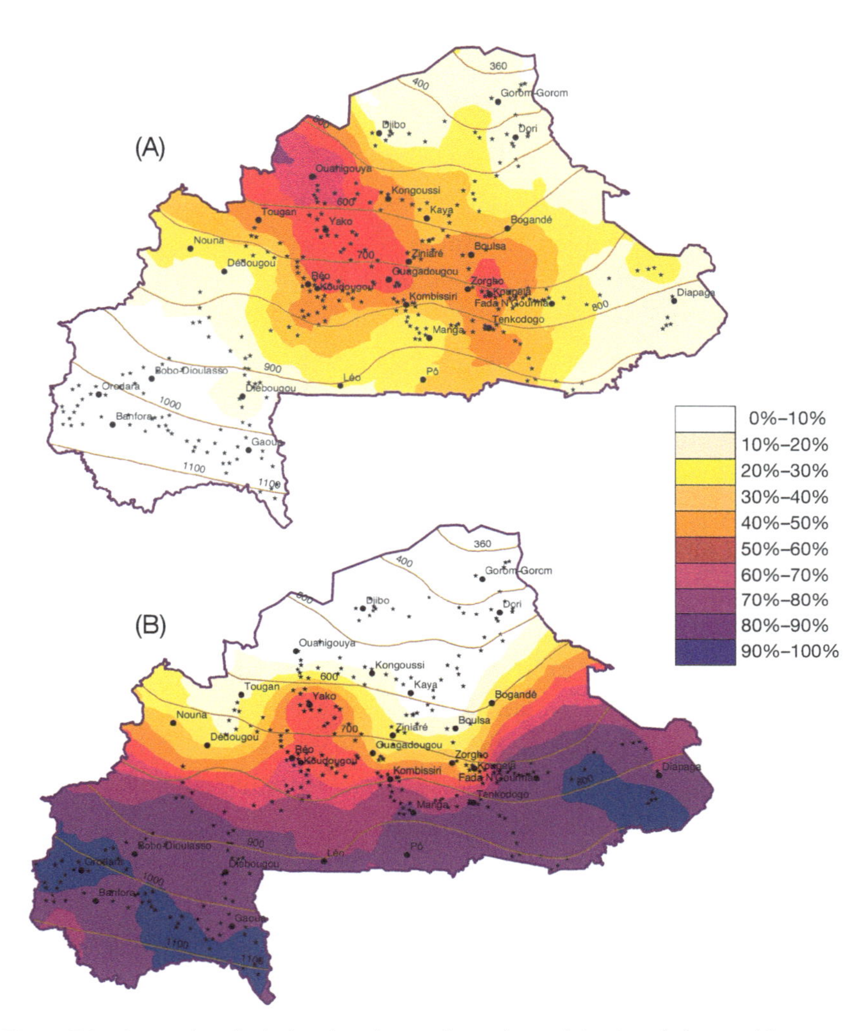

Figure 7.2 Interpolated relative abundance of members of the *Anopheles gambiae* complex in Burkina Faso, based on starred locations on the maps. Dot are cities with >10,000 inhabitants. Lines denote mean annual rainfall isohyets for 1970–2000. **A**. *An. arabiensis* versus *gambiae* s.s. **B**. *An. gambiae* s.s. (molecular form S) versus *coluzzii* (molecular form M). From Costantini et al. (2009).

This has been done in the greatest detail for *gambiae* s.l. in Burkina Faso (Costantini et al. 2009; figure 7.2) and Cameroon (Simard et al. 2009). Three taxa were considered: *arabiensis*, *coluzzii* (M form), and *gambiae* s.s. (S form). In both studies, the geographical variables with the highest association with the presence of all taxa are anthropogenic: roads, villages, and the like. This is not surprising, given that all three taxa are anthropophilic (chapter 6). Flat open areas are generally preferred over more-forested locales. As expected, the presence of *arabiensis* is correlated with increased aridity in locales where *coluzzii* and *gambiae* s.s. are either absent or much lower in abundance, such as in northcentral Burkina Faso and northern Cameroon. Also, the relative abundance of *coluzzii* and *gambiae* s.s. is quite geographically differentiated. *An. gambiae* s.s. tends to prefer warmer, drier habitats, including villages, whereas *coluzzii* tends to be present in moister places with some vegetation. This leads to more overlap between *arabiensis* and *gambiae* s.s. than with *coluzzii*.

Attempts to predict changes in species distributions are closely related to niche modeling, particularly in anticipating the effects of climate change. Two of our mosquitoes, *aegypti* and *quinquefasciatus*, are invasive and clearly temperature limited in their distributions. Furthermore, temperature tolerances by the three genera are somewhat distinct, predicting that they may respond differently to climate changes related to temperature (Couper et al. 2021).

Not surprisingly, *aegypti* is predicted to move into higher latitudes with warming temperatures. Liu-Helmersson et al. (2019) provide more or less typical results for predicted changes over approximately 100 years (from ~2000 to ~2100) under two assumptions about carbon emissions. Importantly in these models, geographic expansion is only moderately affected, primarily in the United States and Australia. An increase in abundance in regions already occupied by *aegypti* appears to be more widespread than geographic expansion, so an increase in the transmission of diseases is more likely (Iwamura et al. 2020).

In the case of *gambiae*, distributional change in Africa is most likely to occur altitudinally—that is, distributions are more temperature limited at high altitudes than at different latitudes. Warming temperatures allow breeding at higher altitudes and, coupled with deforestation, could increase the area occupied by *gambiae* s.l. (Afrane et al. 2012), since most species in this complex prefer open landscapes over forested areas.

An informative study by Giesen et al. (2020) compared the relative impact of climate change on the distribution and intensity of *Anopheles*-borne malaria and *Aedes*-borne viruses in Africa. Moreover, Mordecai et al. (2020) have argued that the major disease burden in Africa could shift from malaria (predominantly *gambiae* transmitted) to arboviruses (*aegypti* transmitted), due to climate change. Brugueras et al. (2020) focus on the impact of climate change on mosquito-borne diseases in southern Europe.

In the case of *quinquefasciatus*, some studies predict only minimal climate change effects on its worldwide distributions (Samy et al. 2016). Relatively cool Canada only has *pipiens* s.s., which is confined primarily to southern regions in eastern Canada. Under models of climate change, the main predicted shift is to the west and, less so, to the north of Canada (Hongoh et al. 2012).

Limits on Climate Change Predictions

Attempts to predict future distributions using climate change modeling need to be viewed with caution, for at least three reasons.

1. Defining the niche of mosquito species with a few variables, mostly related to physical parameters, may not accurately capture essential variables needed to make accurate predictions. Biotic interactions (e.g., competition) may be more important than abiotic variables. This is easily illustrated by models trying to capture the niches of mosquitoes primarily using physical variables. Such models predict that much of Asia and South America *should* have *gambiae* today (C. Li et al. 2021), but they do not. Likewise, much of the Mediterranean area and other warmer parts of Europe *should* have *aegypti* (Wint et al. 2022), but they do not.
2. Predicting how climates will change—and by how much—is also something of a guessing game. Generally, studies use multiple assumptions that are thought to be reasonable, but outcomes can be quite different across "reasonable" models.
3. All modeling assumes that mosquito populations will remain as they are today, rather than evolving in response to climate

change. Couper et al. (2021) make a good case that mosquitoes have a high capacity for developing greater heat tolerance, so in places where high temperatures limit distributions, climate change may not have much impact.

Looking to the Past for Insights

Despite these limitation on *predicting* changes in distributions, there is little doubt that climate change will affect mosquito distributions. The strongest evidence comes from examining what *has already happened*. Carlson et al. (2023) provide an analysis of anophelines in Africa, including *gambiae*. After examining 118 years of records (1898–2016), they conclude that anophelines have increased their range altitudinally by 6.5 m per year and moved toward the poles at 4.7 km per year. These are not trivial changes. And with increasing rates of climate change, whatever the details may be, it is a certainty that—especially for tropical species (*gambiae*, *quinquefasciatus*, and *aegypti*) in places where temperature is the limiting factor—distributions will expand.

In addition to affecting the distribution of mosquito vectors, climate change may also affect the transmission of diseases, due to the effects of temperature and rainfall on the dynamics of transmission. Shocket et al. (2024) provide a recent overview of studies of this sort.

Surviving Harsh Periods

All three mosquitoes face periods of time (usually annually) where severe climates preclude breeding. For the two native African mosquitoes (*aegypti* and *gambiae*), their major challenge is prolonged dry periods, when larval breeding sites dry up. These species rise to the challenge in different ways. *Ae. aegypti* eggs remain viable for some months, while *gambiae* aestivates. For temperate populations of *pipiens* s.s., their Achilles' heel is cold temperatures, which they survive in a state of diapause.

Aedes aegypti

Eggs kept humid but not wet can survive for considerable periods in the lab, often several months (chapter 3). Trpis (1972) made one of the few studies of *aegypti* egg survival in the field. He found up to 40% of the eggs survived for 120 days in a shaded suburban area in Dar es Salaam, Tanzania, but in sunny exposed sites, a maximum of 23% survived for only 60 days. Russell et al. (2001) found considerably lower egg survival rates in Queensland, Australia. This appears to be due to a combination of predation by cockroaches and mold infections. When protected from these factors, their survival rate was 10% after four months.

In contrast to Africa and Queensland, where heat and dryness are the stressors, at higher latitudes, cold can be the limiting factor for *aegypti*. Sol de Majo et al. (2016) studied egg hatching and larval survival in Buenos Aires at temperatures typical of winters there. More than 45% of the eggs hatched, and larval survival to adults was 30% at 13°C.

Another issue is whether all *aegypti* eggs hatch when first flooded with water, or is there "bet hedging," in the sense that only a subset of eggs hatch, with more hatching at subsequent floodings, a process sometimes called "installment hatching." Comparing Aaa and Aaf from Rabai, Kenya, D. Moore (1979) observed proportions of eggs hatching in up to eight rounds of water submersion. A greater portion of Aaa hatched at the first submersion (67%), and 90% had hatched by the eighth. Only about 13% of Aaf eggs hatched at one submersion, and only a cumulative 30% after eight. It is unclear whether this difference is unique to the populations used in this study or represents a trait across these subspecies.

Anopheles gambiae

In West Africa, particularly in the Sahel, prolonged dry periods exist that greatly affect *gambiae* populations. Members of the *gambiae* complex that tend to breed in unforested areas face the extreme challenge of surviving for up to six months with little or no rainfall. It is likely that *gambiae* populations remain *in situ* (i.e., do not travel long distances) and then reestablish themselves when the climate allows. Evidence for this includes the fact that almost as soon as rain appears following the dry period, *gambiae* adults can be captured. There is no time for them to have flown in over hundreds of kilometers, from places where moister conditions prevail year round. Furthermore, at least for *arabiensis*, genetic evidence from the frequencies of inversions over time imply that reasonably large *in situ* populations (N_e) exist throughout the year, including during dry periods (Taylor et al. 1993).

The most direct evidence showing that it is the adult stage of aestivation that allows *coluzzii* to persist through prolonged dry periods in West Africa comes from marking adults prior to the onset of the dry season. Lehmann et al. (2010) found one such female at the onset of the wet season. Faiman et al. (2022) fed a stable isotope, deuterium (^{2}H), to *coluzzii* in villages in Mali that were in the Sahel before the onset of the dry season. After seven months, up to 20% of the captured adults were marked with ^{2}H. It is unclear where these adults "hide," although Lehmann et al. (2014) presented evidence that there were hotspots in villages that had an unusually high density of *coluzzii* adults in the dry season, perhaps indicative of suitable shelters.

Omer and Thompson (1970) found that the majority of *gambiae* s.l. females collected in a nine-month dry season in the Khartoum region of Sudan had fresh or partially digested blood meals. By examining the stage of development of their ovaries, the authors found "ovarian development extremely retarded and only one batch of eggs matured during the whole 9-month period." This is consistent with adults reducing their metabolic activity (aestivating) in response to prolonged drought.

A caveat: In the earlier passive transport discussion, a study was cited where windblown *gambiae* was detected at high altitudes (Huestis et al. 2019), so it is conceivable that *gambiae* populations may also be reestablished or supplemented by long-distance migrants.

Culex pipiens

A third way to survive harsh periods is illustrated by *pipiens* s.s. This temperate-climate member of the complex survives cold winters by diapausing. None of the other members of this complex diapause, even in places where the anautogenous *molestus* form co-occurs with diapausing *pipiens* s.s. (Spielman and Wong 1973).

The onset of cold in temperate zones is accurately predicted by a shortening of day length, which is the major cue for *pipiens* to enter diapause, although falling temperatures enhance the effect of this signal. At day lengths of 15 hours or more, all females remain reproductively active, while day lengths of 12 hours or less induce all females to enter into diapause (Spielman and Wong 1973). The fourth instar larvae are the sensitive stage responding to this cue. Larvae destined to produce diapausing adults spend an extra day as larvae, accumulating lipid (Q. Zhang and Denlinger 2011).

Only females enter diapause, accompanied by several changes. The major one is a cessation of seeking blood meals and an increase in sugar consumption, leading to lipid accumulation (Bowen 1992); blood cannot provide sufficient lipid reserves to survive the winter (Mitchell and Briegel 1989). The ovaries shut down and no egg production occurs (Spielman and Wong 1973). In a study in New York City, Andreadis et al. (2010) provide evidence of the importance of delaying reproduction in order to survive the winter. At the beginning of the cold season in December, 13%–22% of collected females were parous (had produced eggs), but by April, this had dropped to 1%–10%. Similarly, in a locale in southern England, none of the parous females survived the winter, even though 90% of the females in diapause had mated (Onyeka and Boreham 2009). (The longevity of sperm in the spermatheca makes male survival irrelevant in beginning the first spring generation.)

E. Field et al. (2022) offer a more comprehensive study of *pipiens* diapause across the United States. The abundance of nondiapausing females is taken as a proxy for the initiation of diapause, as diapausing female are not attracted to the traps. An important conclusion from their research is that the induction of diapause requires both shortening day lengths and decreasing temperatures. This had been demonstrated in the lab, and these field studies confirmed it. Day length is tied to latitude, and populations at the same latitude (e.g., California and Colorado) differ in the timing of when diapause starts, almost certainly due to lower winter temperatures in Colorado, where diapause begins earlier in the calendar year.

Physiological and molecular changes associated with diapause in *pipiens* include shutting down juvenile hormone production, an increase in body fat, and changes in the expression of genes producing enzymes that are associated with lipid storage (Robich and Denlinger

2005). Denlinger and Armbruster (2014) is a good entrée into a growing body of literature on these aspects of diapause.

Given that some members of the *pipiens* complex diapause while others do not means that the genetic basis of diapause can be studied by hybridizing contrasting phenotypes. Mori et al. (2007) used crosses between *pipiens* and *quinquefasciatus*, which resulted in evidence that the trait is likely highly multigenic. One issue these authors raise is the difficulty of defining a single phenotype for diapause, given the multiple traits associated with this state. Thus it is not surprising that a complex genetic basis is inferred. Meuti et al. (2015) provide evidence of a strong maternal effect. Offspring where the mother was *pipiens* had a much higher proportion of their own offspring being capable of diapause than offspring where the mother was *quinquefasciatus*.

Diapausing *pipiens* females choose sheltered locations (hibernacula) to wait out a cold spell. These may be caves, tunnels, or protected areas near buildings. An MRR study done at a location in New York State (at 43° N latitude) found females marked the previous fall were active in April, providing direct evidence that diapausing females are the founders of the next generation (Ciota et al. 2011). Buffington (1972) noted, “There is extensive movement by most of the population even during midwinter.” In other words, once females are in diapause, they may still move around to some degree.

Predation

Egg Predation

Some of the same predators (fish, tadpoles, and aquatic insect larvae) that consume mosquito larvae may eat mosquito eggs. Egg predation is likely to affect females’ oviposition choices. Vonesh and Blaustein (2010) reviewed 32 published studies relating oviposition choices to the presence/absence of predators. Perhaps not surprisingly, there is no evidence for predator avoidance in *aegypti*, which lays its eggs above the waterline, whereas in both *pipiens* and *gambiae*, which lay their eggs directly on water, there is widespread avoidance of aquatic predators (fish and insect larvae) when ovipositing.

While there is abundant evidence of predators affecting oviposition choices, it is less clear in many studies whether changes in the quantity of eggs and/or the resulting larvae in the presence/absence of predators is due to an oviposition choice or actual predation on eggs. *Culex* egg rafts are particularly attractive to some predators. Segev et al. (2017) provide direct evidence of fish predation on egg rafts in an outdoor mesocosm (close to natural conditions) experiment. Blaustein et al. (2014) found direct evidence for egg raft predation by salamanders. In both these studies, *pipiens* was not included, but it seems safe to extrapolate that, since their egg rafts are almost identical to those of the mosquito species that were studied, *pipiens* would be equally subject to such egg predation.

Larval Predation

Larval predation was discussed in chapter 4.

Adult Predation

Like all flying insects, adult mosquitoes are subject to predation by bats and insectivorous birds. There have been few studies attempting to measure the impact of adult predation on population size, however.

Reiskind and Wund (2009) found a 32% reduction in egg laying by local *Culex* species in outdoor enclosures with bats, compared with sites where bats were excluded. They show that this was not due to a reduction in oviposition, but to predation on ovipositing female adults.

While most adult mosquito predators are insect generalists, with mosquitoes usually composing no more than 10% of the predators' nutrient intake, due to their small size, Africa has a jumping spider (*Evarcha culicivora*, also known as a vampire spider) that predates on blood-fed *gambiae* and *Culex*. Up to 70% of its diet comes from these mosquitoes (Wesolowska and Jackson 2003). This is the only evidence of a predator that is almost solely dependent on mosquitoes.

Ecosystem Services

Do our three mosquitoes play any essential role in the ecosystems of which they are a part? If control measures eliminated or greatly reduced a particular species, would other species be affected? In a thorough review of possible effects of eliminating *aegypti* (and *albopictus*), Bonds et al. (2022) concluded that such control "is likely to be of negligible or limited impact on nontarget predators." The fact that *aegypti* is invasive throughout most of its range means that ecosystems have not had long to evolve a dependence on the species. Similarly, in a review of predation on *gambiae*, Collins et al. (2019) conclude that almost all predators are generalists, and *gambiae* provide only a small part of their diets. While no comparable review has been published on the potential risk of removing members of the *pipiens* complex, it seems unlikely that they would be significantly different from these other two groups.

In addition to providing nutrients for predators, chapter 5 discussed the role of *pipiens* in plant pollination, especially for asters (Peach and Giles 2020). *Ae. aegypti* pollinates some orchids (Lahondere et al. 2020). While members of the *gambiae* complex are known to visit plants to obtain nutrients, there is little evidence that they affect pollination.

Notes

1. The Breteau Index (the number of positive containers per 100 houses) is the most widely used way to express relative *aegypti* numbers. "Containers" may be larval sites or ovitraps. Nathan et al. (2006) offer a critical assessment of this index.

2. In addition to being regulated by local governments, some funding agencies have restrictions on this method of collecting if the research is being supported by them. For the United States' National Institutes of Health, approval by the institutional review board of the parent organization is required, as well as compliance with government regulations in the country where the collecting is to be done. Because HLC is most often used for *gambiae*, malaria is the disease of greatest concern. Providing collectors with antimalarial medication during the period when they will be exposing themselves greatly reduces the risk of them contracting malaria (Gimnig et al. 2013). Most of the major diseases transmitted by the other two species (*aegypti* and *pipiens*) do not have such prophylactic medications. Achee et al. (2015) and Benedict et al. (2018) provide more discussions on the use of humans in vector biology research.

3. Studies were done over time, and in some cases changes occurred in the release locations, as well as in the placement of traps for recapture. Not all studies allow researchers to separate out each release by time, locality, and trap arrangement.

4. The particular species in each of these studies varied. In Mali, both *gambiae* s.s. and *arabiensis* were present. In Kenya, only *gambiae* s.s. was present, while in Burkina Faso, results were presented only for *gambiae* s.l.

5. M. C. Crosby's 2006 book, *The American Plague*, is a highly readable and (dare one say) entertaining account of this short-lived epidemic.

6. When *gambiae* was initially found in 1930, it was still considered a single species (chapter 1), so it was not known which of the morphologically identical species was introduced. Using museum specimens collected in Natal at that time, sufficient DNA sequencing data were obtained to confirm that it was *arabiensis* (Parmakelis et al. 2008). This was perhaps to be expected, as *arabiensis* is the most arid adapted of the taxa in the *gambiae* complex, and this region of Brazil is very dry.

7. This interspecific hybridization between *albopictus* and *aegypti* has been dubbed "satyrization," named for the Greek god Satyr, a quasihuman mythical creature (Ribeiro and Spielman 1986). Male satyrs mate indiscriminately, especially with female nymphs. There is precious little of what might be called taxonomy in Greek and Roman mythology, so it is not clear if nymphs and satyrs represented different species.

CHAPTER 8

Genetic Variation

> Uniformity is not nature's way;
> diversity is nature's way.
> —Vandana Shiva

> I'm one of those people you hate
> because of genetics. It's the truth.
> —Brad Pitt

Typology

As the then-new field of genetics was developing in the first half of the 20th century, an important concept was the "wild type," indicating the normal or typical form of a species. Wild type homozygotes are the norm, and mutations are deviations from the predominate wild type. This was a useful paradigm, as it led to great advances in demonstrating that mutations behaved as Mendel predicted: they could be mapped into linear arrays, assigned to chromosomes, and the like. Thomas Hunt Morgan's laboratories at Columbia University and then the California Institute of Technology were the early centers of advances in understanding these basic aspects of inheritance.

The implications of these advances for understanding evolution were quickly realized, and attempts to incorporate Mendelian genetic inheritance into evolutionary theory led, in the 1920s to the 1950s, to the development of population genetics. Provine (1971) is an excellent introduction to the history of this fundamental advance in biology. Yet an apparently irreconcilable issue soon arose in attempts to develop what became known as a synthesis of Darwinian evolutionary theory with the new field of genetics. Darwinian evolution (by natural selection) requires heritable variations for an assortment of traits in populations found in nature, whereas genetic work defined wild type as the dominant genotype, with variants (mutations) being rare, usually arising as new mutations. These two views of genetic variation in populations became labeled "typological" (from Plato's *eidos*, meaning an essential type) and "populational" thinking. The former applies to the Morgan school, which emphasizes the uniformity of populations. The latter posits that species are not dominated by a single genotype, but instead compose populations with a large number of genotypes.

As data on the genetics of natural populations accumulated through the 20th century, the populational view became dominant in evolutionary biology, as it became clear that genetic variation abounds in populations. Lewontin's 1974 masterful book, *The Genetic Basis of Evolutionary Change*, reviews evidence for the conclusion that natural populations harbor a plethora of genetic variation.

These diametrically opposed views are relevant to mosquito biology and have affected research agendas into the present (Powell 2018a). Overwhelming empirical evidence supports populational thinking for vectors, especially mosquitoes. This is crucial in understanding how these insects affect human health. It also potentially shows how mosquitoes can be controlled to reduce human suffering. Selection for naturally occurring genetic variants rendering females incapable of disease transmission holds out one of the best hopes of reducing human suffering from mosquito-borne diseases (Powell and Tabachnick 2014; Xia et al. 2019).

Morphological Variation

Like classical genetics, initial genetic work on mosquitoes involved finding visually recognizable mutations and determining their modes of inheritance, linkage, and the like. Not surprisingly, given the ease with which it can be cultured in the laboratory, *Ae. aegypti* played a central role in early mosquito genetics.

Aedes aegypti

Some 30 or so morphologically distinguishable mutations were characterized in *aegypti*, including mapping them to the mosquito's three chromosomes (Munstermann and Craig 1979; Munstermann 1994). These include eye color variants, proboscipedia (leg-like proboscis), and a miniature body, similar to well-known mutations in *Drosophila*.[1] Some of these mutants can be found in natural populations (e.g., Verna and Munstermann 2011).

By far the most ubiquitous and important morphological variation in *aegypti* applies to scale patterns and cuticle coloration. McClelland (1974) provides an extensive treatment of this variation in collections from across the range of *aegypti*. Figure 1.2 illustrates some of his findings. The two subspecies of *aegypti* (Aaa and Aaf) were initially described and defined by their degree of white or silver scaling (chapter 1). As is clear from McClelland's classic study, however, the degree of white scaling is a more or less continuous trait that only roughly follows geographic distributions. Not surprisingly, the genetic basis for this variation is complex, involving several genes and modifiers. Fully black Aaf exist only in Africa, whereas *aegypti* with some white scaling have a mixed distribution. The *mean* degree of white scaling decreases when going from the Caribbean (including Florida), where *aegypti* has the greatest degree of white, to West Africa, where many populations lack any abdominal white scales. Using only these scaling patterns to define Aaf and Aaa, however, has led to confusion in the literature, and caution needs to be applied when assigning subspecific names to populations. Jupp et al. (1991) reared offspring from a single wild-collected female that varied across almost the entire range of scaling, which further documents the danger of using this phenotype to assign subspecific names to specimens. Despite this geographic heterogeneity in overall white scaling and abdominal coloration, if one considers only the dorsal pattern on the first abdominal tergite, the degree of scaling does quite closely follow the genetic distinction between Aaa and Aaf determined by DNA sequencing data (figure 3 in Rose et al. 2020).

Anopheles gambiae

Morphologically, adults in this complex are remarkably uniform. Relatively few morphological mutations have been found or studied within any species of the complex (Kitzmiller and Mason 1967), while more effort was put into morphologically distinguishing sibling species. In a painstaking study, Coluzzi (1964) examined up to 350 different morphological characters in 3,200 specimens from 29 populations, including eggs, larvae, pupae, and adults. Both saltwater breeders (*merus* and *melas*) and species A, B, and C (chapter 1) were studied. He concluded that "morphological differences may be non-existent." The only reliable morphology that distinguished species were the pecten (spicules, or spines) on the siphon of the larvae in the saltwater species, compared with the freshwater breeders. Presumably this is an adaptation to salinity.

Culex pipiens

Laven (1967) lists some 20 morphological mutants found in *pipiens*, some of which are single genes, while others are multigenic. More intriguing are the subtle morphological differences that have been used to distinguish members of this complex. Quantitative differences in the ratio of two male genital structures distinguish *pipiens* from *quinquefasciatus* (Barr 1957).

Chromosomal Variation

In addition to morphological and molecular genetic diversity, caused by nucleotide substitutions, many insects, including mosquitoes, have important variations in the order or position of nucleotide sequences—that is, variations in their chromosome structure. Inversions of blocks of genes are the most widespread, naturally occurring type of chromosomal variation. The significance of chromosomal inversions in understanding many aspects of biology stems from extensive work on *Drosophila*, with favorable polytene chromosomes allowing the detection of chromosomal variants. Krimbas and Powell (1992) provide a compilation of this kind of work in *Drosophila*.

The early detection of inversions in *Drosophila* was due to their effect on recombination. Strains were found with "recombination suppression" in parts of the genome, and these suppressors were shown to be inverted chromosomes. Crossing over between breakpoints in inversion heterozygotes may occur, but the resulting chromatids are unbalanced (i.e., have duplications and deletions) and cannot produce functional gametes, and thus are not represented in offspring.[2] This means that entire blocks of genes between breakpoints are inherited intact, as single Mendelian factors. These have been dubbed "supergenes," indicating that they consist of many genes but are inherited as a single block (Kirkpatrick 2010; see also critiques in Villoutreix et al. 2021). The importance of these intact blocks of genes is that they are often subject to strong selection and consist of co-adapted alleles at multiple loci, thus conferring selective advantages. As will become clear, at least in the *gambiae* complex, chromosomal inversions play a fundamental role in many aspects of the biology of these mosquitoes (chapters 9 and 10).

Aedes aegypti

Tissues in *aegypti* do not have favorable polytene chromosomes suitable for easy examination of chromosomal variation (chapter 2). Until recently, there was limited information on the possible presence of inversions in *aegypti*, and what little information does exist is seemingly contradictory.

Bernhardt et al. (2009) crossed Aaf from Senegal with Aaa (either the ROCK strain or one from Puerto Rico), producing offspring that showed greatly reduced recombination, and linkage groups that were at odds with established gene orders. The authors inferred that at least four inversions on chromosome 1, one inversion on chromosome 2, and three inversions on chromosome 3 were causing the results they observed. Dickson et al. (2016) studied the same crosses and used FISH (fluorescence *in situ* hybridization) to physically distinguish the gene order on the chromosomes. As their mapping studies indicated, inversions were detected in all chromosomes.

In addition to direct microscopic observation that can reveal relatively large chromosomal variants, DNA sequencing of genomes can also disclose smaller inversions. Redman et al. (2020) used DNA sequencing techniques to detect changes in the gene order for two Aaf strains (from Uganda and Gabon) and two Aaa (from New Orleans and Thailand), as well as a hybrid zone in eastern Kenya (Rabai). They detected no evidence for large inversions of the sort inferred from the crosses using Senegal strains (discussed above). They did, however, detect 32 instances of what they called "microinversions" (median size = 43 kb) that were widespread in both the Aaa and Aaf strains.

The most convincing study of inversions in *aegypti* is by Liang et al. (2024). They used both a DNA sequencing method (Hi-C proximity ligation) that can detect changes in the physical location of specific sequences that would result from inverted chromosomes, followed by microscope mapping using FISH. They studied 23 recently collected strains from across the distribution of the species, as well as two common laboratory strains and *Ae. mascarensis*. Figure 8.1 summarizes their findings. While this study found more inversions in African strains than in those from outside Africa and no inversion was found in both regions (Aaf and Aaa), their sampling was quite limited. Further sampling may change this conclusion.

It remains to be unambiguously determined whether naturally occurring chromosomal inversions are sufficiently common in *aegypti* to significantly affect population genetic dynamics and analyses. Virtually all population genetics work done on *aegypti* has assumed that there were no inversions—that is, loci are inherited independently. This assumption has not led to any obvious contradictions or problems in the interpretation of data. Furthermore, examination of whole genome sequences (WGS) for 1,200 *aegypti* genomes has not revealed any blocks of high LD, as would be expected if inversions were common and suppressed recombination (Crawford et al. 2024).

Anopheles gambiae

The *gambiae* complex has abundant, naturally occurring chromosomal inversion polymorphisms on all chromosomes, as well as in all sibling species. Because highly favorable polytene chromosomes can be prepared

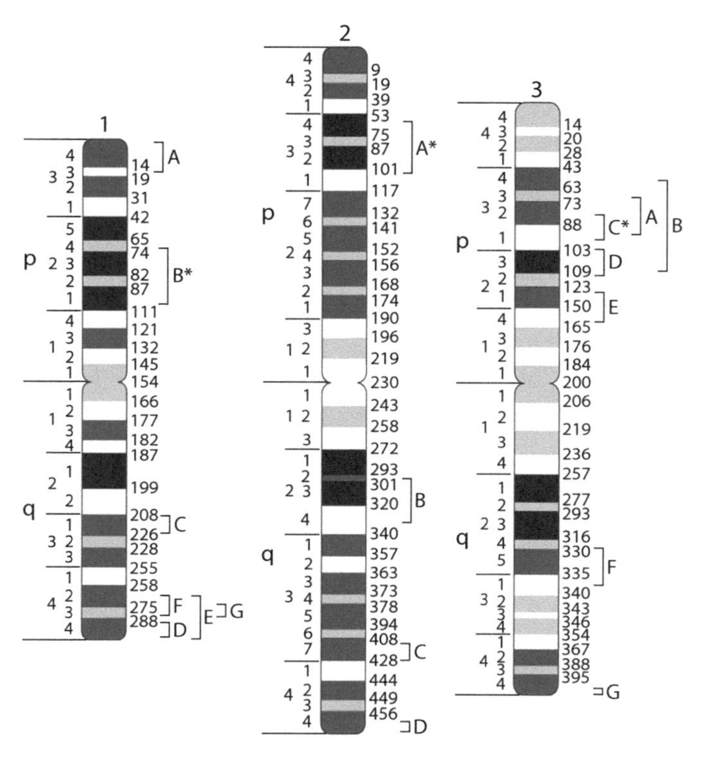

Figure 8.1 Chromosomal map of inversions in *Aedes aegypti*. Short and long arms are indicated as p and q, respectively. Numbers on the left indicate chromosome regions, and numbers on the right show genome coordinates in Mb. Inversions found in Africa are in regular typeface, and those found outside Africa have an asterisk. From Liang et al. (2024).

from these species, inversions have been extensively studied and play crucial roles in the biology of the complex. (Figure 2.4 presents photos of inversion heterozygotes.) In *gambiae* s.s., 2R (right arm of chromosome 2) is particularly polymorphic (figure 8.2). While this number of inversions might be difficult to study in populations, in fact only five and one are common in 2R and 2L, respectively (denoted by brackets under the arm drawings in figure 8.2). Also note that these common inversions are paracentric, i.e., do not include the centromere.

The importance of inversion polymorphisms in the *gambiae* complex will be discussed in some detail in chapters 9 and 10.

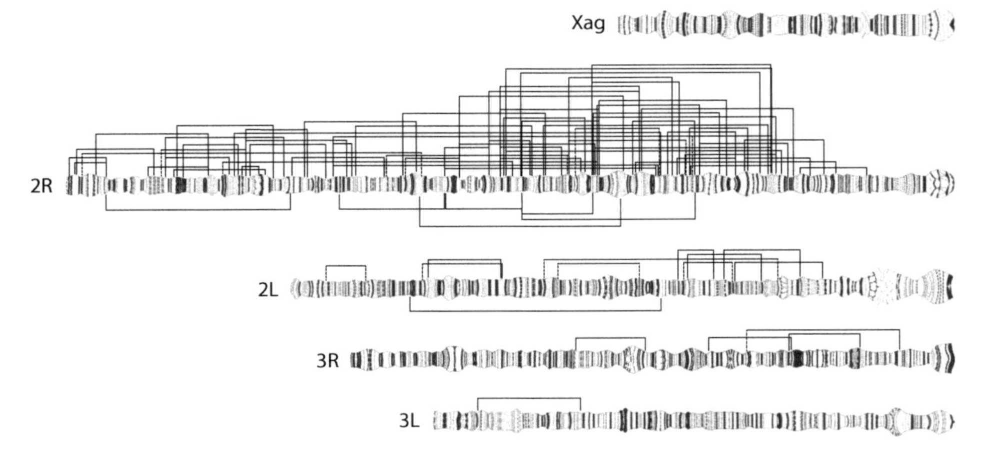

Figure 8.2 Naturally occurring paracentric inversions in *Anopheles gambiae* s.s. Brackets above the chromosomes are rare inversions, while those below indicate common ones, as defined in Pombi et al. (2008).

Culex pipiens

There are a few reports of naturally occurring inversions in the *pipiens* complex. Tewfik and Barr (1976) describe an inversion on chromosome 2 that involves 18% of the length of the arm. McAbee et al. (2007) found two inversions in chromosome 2 of *C. quinquefasciatus*. Similarly, Unger et al. (2015) found two inversions, one in chromosome 2 and one in chromosome 3, covering, respectively, about 15% and 10% of the arm involved. This publication has photographs of the most informative polytene chromosomes made for this complex. McAbee et al. (2007) report a success rate of only about 20% in obtaining usable chromosomes from families of field-caught *quinquefasciatus*, and similar high-quality preparations were not obtained for *pipiens* s.s. This precludes routine population and ecological genetic work of the sort possible for *gambiae*.

Molecular Variation

How much genetic variation is there in natural populations? This is a deceptively simple question that has been definitively answered only after extensive DNA sequencing became feasible. Before whole genome sequencing became widely used, indirect methods of determining levels of variation were employed (Powell 1994). While used less often today, their application to our three mosquitoes provides important information worth summarizing.

The first method detected variation in amino acid sequences in proteins, followed later by a direct examination of DNA sequences. Extensive allozyme studies were performed on these three mosquitoes,[3] especially *aegypti*. Indirect methods to assay variation at the DNA nucleotide level were then employed. Examples of the use of one such method, RFLP (restriction fragment length polymorphisms) on *pipiens*, will be presented.

Because the vast majority of measurements for genetic variation in populations today are through direct DNA sequencing data or microsatellites,[4] the application of these methods will be presented more fully both below and in chapters 9 and 10.

Aedes aegypti

The first extensive study of allozyme variations in a mosquito was for *aegypti*, done largely in the author's laboratory (Tabachnick and Powell 1978; Wallis et al. 1984). One of the most important results from these studies was the documentation of extensive genetic variations that were very highly geographically structured (Powell et al. 1980), a harbinger of what has been found with newer methods.

A large microsatellite database for *aegypti* exists (chapter 9). For the present chapter's focus on levels of genetic variation, estimates of the diversity observed, summarizing heterozygosity and allelic richness for 11 microsatellite loci are given in table 8.1. It is clear that these loci are highly polymorphic. Allele number varies from about 5 to 11 in different regions, with ancestral African populations having more alleles. Similarly, numbers for "private" alleles (alleles found in only one sample) are greater in Africa. This pattern is completely consistent with the hypothesis that African populations are older than those outside Africa and thus have had time to accumulate more genetic diversity. Overall, microsatellite data from *aegypti* are consistent with the earlier allozyme work, although they provide much clearer resolution, both because the

Table 8.1 Microsatellite data for worldwide samples of *aegypti*

Region	N	H_O	AR (200)	PAR (200)	F_{ST} Intra
Africa	918	0.591	10.97	3.41	0.114
N. America	952	0.548	6.36	0.14	0.149
S. America	562	0.489	7.37	0.30	0.181
Asia	603	0.551	7.87	0.58	0.121
Pacific	73	0.557	4.75	0	0.123

Note: N is sample size; H_O is observed heterozygosity; AR and PAR are allelic richness and private alleles, standardized by rarefaction (N = 200 genes); F_{ST} Intra is the average F_{ST} between populations. Data from Gloria-Soria et al. (2016a).

data are more informative, and the methods to analyze such data have greatly improved. The relative power of various types of population genetic data is further illustrated in chapter 9.

Direct studies of DNA nucleotide variation (SNPs, single nucleotide polymorphisms) in *aegypti* include a variety of technologies, the most informative of which are RAD-seq, SNP chip, and WGS. These will be discussed in the next chapter, so for now I simply present the estimates of sequence variations, based on whole genome sequences. One common way to quantify the level of genetic variation from DNA sequencing data is designated π, for the mean number of nucleotide differences between two randomly sampled haplotypes in a population. Figure 8.3 shows estimates for π across the genome in a worldwide sample of populations. Three features of these data are worth emphasizing and are discussed in detail in Crawford et al. (2024).

1. Variation across the genome is greatest in the middle of the arms and decreases toward the centromeres, likely caused by reduced recombination in centromeric regions. Begun and Aquadro (1992) were the first to describe this pattern, which has been confirmed in virtually every species examined since.
2. African samples are more genetically diverse than those outside Africa.
3. East African samples are more diverse than West African ones,

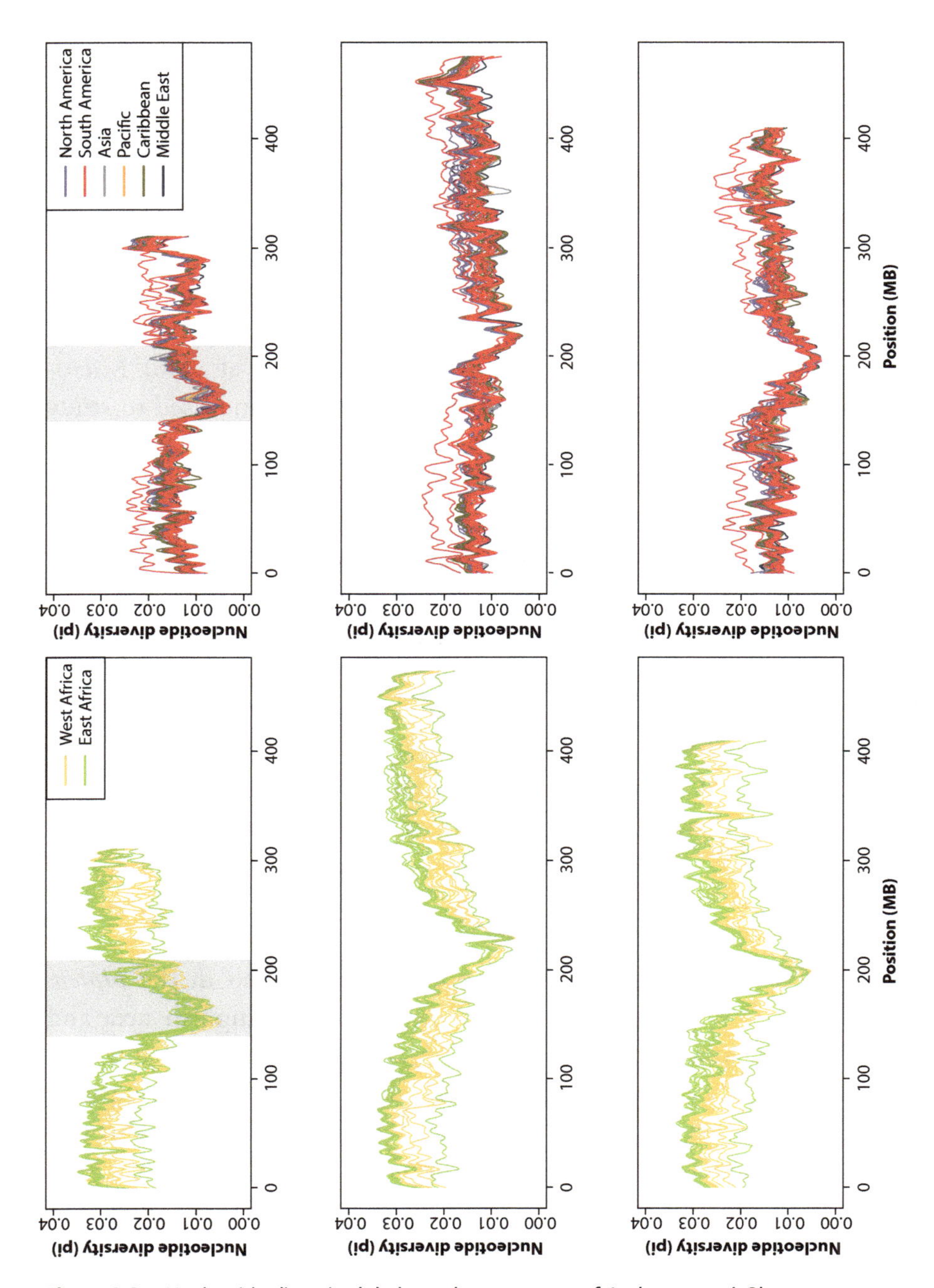

Figure 8.3 Nucleotide diversity (π) along chromosomes of *Aedes aegypti*. Chromosomes 1, 2, and 3 are shown from top to bottom (with page rotated 90° clockwise). Note the decreases at the centromeres. **Right**. Graphs are for samples outside Africa. **Left**. Graphs are for samples from Africa, with green shading for East Africa and yellow for West Africa. The African estimates are greater than those outside Africa. Data from Crawford et al. (2024).

consistent with islands off the coast of East Africa being the origin of continental *aegypti* (chapter 1).

Anopheles gambiae

Allozyme studies on this complex have included efforts to distinguish morphologically identical species (e.g., Mahon et al. 1976) and to elucidate genetic differentiation across Africa (Lehmann et al. 1996). Microsatellites have also been developed for this complex and used to study geographic variation (Lanzaro et al. 1995, 1998), differentiation across the Rift Valley (L. Field et al. 1999), and fine scale house-to-house differentiation within a village in Kenya (Lehmann et al. 1997).

Extensive WGS is the most revealing method with regard to the overall genetic diversity of the *gambiae* complex. Diversity across the genome of six of the members in the complex is very similar to that observed in *aegypti*, with the centromeric regions having a significant dip in SNP diversity (Figure S9 in Fontaine et al. (2015). The two saltwater breeders (*melas* and *merus*) appear less variable than the other species, particularly notable in the X chromosome. Figure 8.4 summarizes the average genomic π of 765 *gambiae* and *coluzzii* specimens across Africa.

Culex pipiens

Allozymes and microsatellites have also been studied in the *pipiens* complex, both for population genetics and to distinguish taxa and analyze hybridization (e.g., Fonseca et al. 2004; Weitzel et al. 2009; Becker et al. 2012; Kothera et al. 2012). Like the other two mosquitoes, genetic variation abounds in *pipiens*. Microsatellite variation in this complex is high, which is true for most mosquitoes. For example, Keyghobadi et al. (2006) observed between 7 and 19 alleles per locus and heterozygosity of 66%–93% in one sample of 29 mosquitoes from New York State.

The use of allozyme analyses for purposes other than population genetics is nicely illustrated by Severson et al. (1993, 1995) for *pipiens*. These workers used this method to construct linkage maps and perform quantitative trait mapping in *pipiens*.

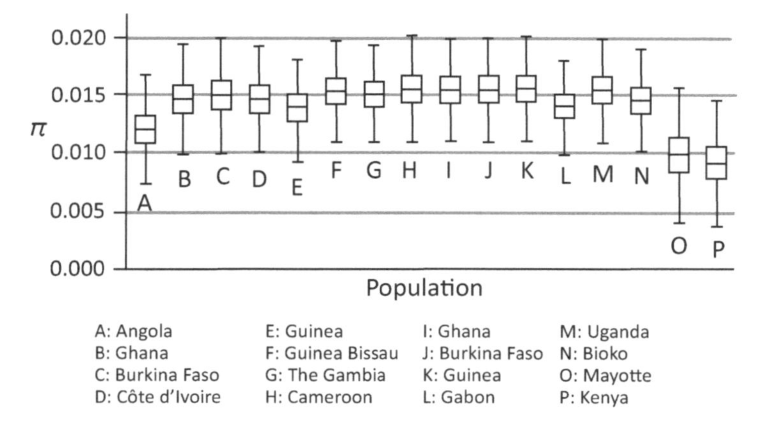

Figure 8.4 Nucleotide diversity in different populations of *Anopheles gambiae* across Africa. Sample sizes for each population averaged 85 individuals, with a minimum of 31 and a maximum 275. From *Anopheles gambiae* 1000 Genomes Consortium (2020).

WGS across three taxa of the *pipiens* complex has been obtained more recently (Haba et al., submitted). The observed patterns are remarkably similar to those of the other two mosquitoes. Overall π is around 1%–3% and decreases near the centromeres.

Conclusion from Molecular Variation Studies

Given that the average heterozygosity per nucleotide site (π) is 1%–3% for these mosquitoes (figures 8.3 and 8.4), this means that two randomly chosen 1,000 bp (1 Kb) segments differ at 10–30 positions. The average protein-coding gene is about 1 Kb (figure 2.2). Protein-coding regions, however, are less variable than the average over the genome—on the order of 40% of the average π. What this means is that, on average in a single population, two randomly chosen alleles of a protein-coding gene differ by at least 4 bp. This implies that *every individual is heterozygous at virtually every gene.*

While this should have ended the typological versus populational controversy discussed at the start of this chapter in favor of populational thinking, it did not entirely do so. Proponents of typological thinking made the very reasonable objection that while molecular studies uncover considerable variation, it is not at all clear how much of this diversity affects the phenotype. From a functional point of view, it may make virtually no difference whether a particular nucleotide is at a particular position in the genome (e.g., variation at the third position of codons that code for the same amino acid, or synonymous variation). This has been called the "neutral theory," positing that the vast majority of genetic variation at the molecular level does not affect function and thus cannot affect the fitness of the carriers. In other words, variation is neutral with respect to selection. It is beyond the scope of this book to address this fundamental issue in any detail. Today, most evolutionary geneticists do not adhere to either extreme view. Some molecular variation must account for adaptations, at least at some point in time in some populations, while for much of the time, the majority of nucleotide variation is effectively neutral. The assumption of neutrality (genetic drift is the only factor changing allele frequency) underlies many of the analytical methods in population genetics (e.g., inferring demographics). In addition, molecular phylogenetics relies on the assumption (rarely explicitly stated) that DNA divergence proceeds unidirectionally, in a more or less clocklike manner, as predicted by random genetic drift.

Elsewhere, I've discussed how these observations affect research agendas on vectors of diseases, including mosquitoes, and noted what they imply for attempts to control vector populations (Powell 2018a, 2022).

Notes

1. George B. Craig Jr., at the University of Notre Dame, was the pioneer in *aegypti* genetics. For many years, he maintained a WHO-sponsored collection of mosquito strains carrying various mutations. Unfortunately, virtually none of these strains exist today.

2. The terms "crossing over" and "recombination" refer to somewhat different processes. Crossing over is the physical breaking and rejoining of homologous chromosomes (chromatids in the four-strand stage of meiosis). Recombination refers to the production of chromosomes in offspring, consisting of a mixture of the two parental homologous chromosomes. Crossing over is the most common cause of recombination, although other processes, like gene conversion, can shuffle genes between homologous chromatids. One might deduce that inversions would be partially sterile if crossing over

between breakpoints results in nonfunctional gametes. But this is not always the case, at least in females. Only one of the four products from meiotic divisions produces an egg, and the other three products are nutrient sources for the egg (nurse cells). In *Drosophila* (and likely mosquitoes), chromatids with one copy of all genes preferentially (nonrecombinants), if not exclusively, form a functional egg, while the unbalanced chromatids are relegated to nurse cells.

3. Because proteins have a net electrical charge (depending on their pH), when placed in a porous medium, across which an electrical charge is then sent, proteins migrate through the medium at a rate dependent on their size and overall charge. This is called "electrophoresis." The first method of detecting molecular variation in populations was the electrophoresis of proteins, visualized by staining to reveal enzyme activity. When more than one protein was found to have the same enzymatic activity, they were called "isozymes." Later, when this method was applied to studies of variation in populations, the relevant isozymes were alleles at the same locus, and "allelic isozymes" became shortened to "allozymes."

4. Microsatellites are regions of genomes that contain many tandem copies of short base-pair repeats, generally 2–13 bp long. The number of these repeats is highly variable across genomes and can be detected by changes in the size of DNA fragments when the repeat unit is amplified by PCR. Generally 5–20 alleles, differing in the number of repeat units, occur in natural populations. The high number of alleles at each locus is particularly powerful in some population genetics analytical procedures. While DNA sequencing provides data on lots of loci, a maximum of only four alleles are possible at any one SNP locus, making microsatellites more informative on a per locus basis.

CHAPTER 9

Population Genetics

> Population genetics has come to occupy a rather special place in biology. It made possible the great renaissance of evolutionary biology which began about 1930. Population genetics then tended to reunite fields of biology such as genetics, ecology, paleontology, and systematics, which had tended to take separate paths. It has thus been referred to, and with cause, as the core subject of general biology.
>
> —L. C. Dunn, 1965

Chapter 8 documented the high level of genetic variation in populations of these three mosquitoes. In this chapter, the issue addressed is how patterns of genetic diversity in populations can illuminate aspects of their biology. Extracting *patterns* in allele frequencies in populations to infer *processes* and *history* is the purview of population genetics. Traditionally, population genetics stopped at the species boundary, as interspecies genetic analyses were not possible. With today's molecular methodologies, researchers can now study and rigorously connect patterns of genetic diversity within species to multiple related taxa. Perhaps another term, like "evolutionary genetics," better describes this advance. In this chapter, I restrict the discussion (with some exceptions) to within-species (single taxon) genetic patterns and leave between-taxa considerations to chapter 10, as such data lend insights into the speciation process.

Nature of the Data

Stating that both population and evolutionary genetics use allele frequencies to make inferences about processes requires a broad definition of allele. This could refer to chromosomal inversions; proteins with different electrophoretic charges (allozymes); frequencies of different-sized segments of DNA due to insertion/deletions (microsatellites); or single nucleotide differences (SNPs). Each of our three mosquitoes have been studied to a greater or lesser extent for each of these types of genetic variation, although only *gambiae* s.l. has been extensively examined for chromosomal variation.

A large part of the history of population genetics is the history of the development of increasingly accurate methods to detect genetic variation in populations. As methods became better and more precise, and their implementation on the population level became easier (e.g., mechanized), older methods were abandoned. Analytical tools advanced hand in hand with advances in data collection. Today, two types of genetic variation are studied most widely.

Tandem Repeats and Length Variation

Virtually all eukaryotic genomes have multiple stretches of nucleotide sequences, called microsatellites (chapter 8, note 4), that have units of 2–13 base pairs that are tandemly repeated 5–50 times. The number of repeat units at any locus is highly mutable, so such loci may have up to 20 alleles of different sizes segregating in a single population. Microsatellites are also known as VNTRs (variable number tandem repeats), SSRs (simple sequence repeats), and STRs (simple tandem repeats). Variation at such loci is detected by designing PCR primers that flank the repeated section, yielding different-sized products (detected by electrophoresis), depending on the number of repeat units between the primers. The highly multiallelic nature of microsatellites provides powerful data for population genetics. The ubiquity, ease, and low cost of microsatellite analyses make them popular in empirical population genetics, including with our three mosquitoes.

DNA Sequences

With improvements in the speed and cost of determining the nucleotide sequence of eukaryotic genomes, whole genome sequencing has become popular in population genetics. One requirement in using WGS data is to have a genome assembly onto which variants found in a sample of individuals can be mapped. All three of our mosquitoes have reasonably good genome assemblies. Their total genome size matches those measured by methods such as flow cytometry, with BUSCO scores of 96%–98% (95% or higher is generally considered excellent).[1] Sequences have been assigned to three chromosomes. Recent publications for the three mosquitoes include ones for *aegypti* (B. Matthews et al. 2018; Morinaga et al., submitted), *pipiens* (Liu et al. 2023; Ryazansky et al. 2024), and *gambiae* (Zamyatin et al. 2021). Assemblies are regularly updated and improved, however, and an online search or consultation with an active worker in the field can provide the latest updates.

Because WGS is expensive and requires considerable computing expertise—and capacity—to analyze population samples numbering in the hundreds, streamlined methods have been developed to produce almost as much information as WGS. One approach is capture sequencing, where a subset of conserved sequences is amplified and sequenced. This is especially efficient when performing phylogenetic analyses across a broad number of species. Soghigian et al. (2023) used capture sequencing of more than 700 genes with sequences of more than 1.5 Mb to obtain well-supported phylogenetic trees across Culicidae (figure 6.1).

Another efficient way to gather SNP data on many individuals is to use a SNP chip. This is a method that employs an array of probes that recognize one of two alternative nucleotides at a site. Chips can be made that assay 50,000–500,000 SNPs—more than enough data for detailed population genetics analyses. Chips are available for *aegypti* for 50,000 SNPs (Evans et al. 2015) and *gambiae* for 400,000 SNPs (Neafsy et al. 2010). One drawback of SNP chip data is that only two alleles at a locus are detected, precluding some types of analyses.

Analyses

A large part of any population genetics study involves analyses, which are sometimes simple, but often complex, especially when large data sets have been generated. The type of analysis that is done is also dependent on the question an investigator wants answered by the data.

Some of these analytical methods are familiar to most biologists, such as principal components analysis and tree joining (based on genetic affinities). Two other methods, both very powerful through using the kind of multilocus data generated by modern studies, are less familiar and will be briefly introduced here.

STRUCTURE and ADMIXTURE

Pritchard et al. (2000) introduced what has become a widely used method, due to its statistical rigor and easily understood visual display. This is commonly referred to as STRUCTURE, generating STRUCTURE plots.[2] One strength of this approach is that a user does not assign individuals in the samples to a particular population. Rather, membership in a genetic subunit found in the data is determined by Bayesian statistical methods. This allows the identification of cryptic subunits a user may not have identified beforehand, as well as the opposite, where samples that were thought to be genetically distinct are not. The investigator only enters the number of subunits (K) into which the program will divide the data, using maximum likelihood for each individual's membership into one of the K subunits. By using different Ks, aspects of the data become clearer, as well as there being a maximum K, beyond which further division of the data adds no more resolution (Evanno et al. 2005). The proportion of loci assigning an individual to each subunit is indicated by the height of various bars in the graphs, with different subunits being color coded, a modification introduced by Tang et al. (2005) and implemented as ADMIXTURE. Lawson et al. (2018) provide advice in interpreting such plots and point out their limitations.

Figure 9.11 (discussed in more detail later) shows a clear distinction between *pipiens* s.s. and *molestus*, as indicated by the different colors. In the upper graph, there is some admixture of *molestus* genes into *pipiens* s.s., but not from *pipiens* s.s. into *molestus*. As K increases, more substructure can be seen in the data, with *molestus* cleanly separating into three distinct groups (corresponding to the collection locales), with no indication of mixing. *C. pipiens* s.s. data also indicate further substructures, but with mixing.

History

Explicitly historical reconstructions from multilocus data include approximate Bayesian computation, or ABC (Beaumont et al. 2002). The data allow several trees that relate the samples to one another to be tested for their likelihoods (i.e., maximum likelihood). It then takes the most likely tree and uses Bayesian methods to calculate how many generations would have passed since the populations split. Figure 9.4A is an example of using microsatellites on *aegypti* in the United States. There is strong support ($p = 99\%$) for the branching scenario shown, with alternatives having minimal support. Given this scenario, the times associated with the nodes are estimated, which requires estimates of N_e (the effective population size). Several N_es are tested, and the one providing the highest probability is chosen. In this example, T1 has a mean estimate of 224 generations, with a lower (2.5%) bound of 40 generations, and an upper (97.5%) bound of 774 generations.

Other common methods to statistically estimate trees and times of divergence are implemented in BEAST (Bouckaert et al. 2014). TreeMix (Pickrell and Pritchard 2012) allows mixtures between lineages (introgression).

Allele Frequency Spectra (AFS) and Coalescence

A complementary analytical procedure developed to extract historical demographic information from multilocus genotype data is based on the array of allele frequencies at many loci (Gutenkunst et al. 2009; Excoffier et al. 2013). Population bottlenecks, expansions, migrations, and the like can be detected. Crawford et al. (2017) gives an example of this for *aegypti*. Most recently, demographic parameters and the timing of events have been estimated using a cross-coalescence approach, such as in figure 9.4B.

What Is Known

Aedes aegypti

Urdoneta-Marquez and Failloux (2011) offer a thorough review of population genetics work done on *aegypti* up to about 2010 and provide a

Table 9.1 Population genetics databases available for *aegypti* as of summer 2023

	Microsatellites[1]	SNP chip[2]	WGS[3]
Population samples	320	232	91
Countries	44	54	35
Continents	6	6	6
Total individuals	~11,000	5,795	1,371

1. All data collected in Yale lab or standardized with Yale allele calls.
2. All data from the same chip processed at the same genome center.
3. Illumina sequencing to 20–30X. From Crawford et al. (2024).

guide into this literature. Here I will present more-recent data, primarily from microsatellites, SNP chip, and DNA sequencing work.[3] Table 9.1 shows the size of available databases for *aegypti* for these three types of data.

Discriminating Subspecies

Figure 9.1A shows data from three different methods of addressing the issue of how genetically distinct Aaa and Aaf are. All methods provide very similar results. Aaa and Aaf are sufficiently differentiated that in most cases assignment can be made using microsatellites, SNP data from a chip, or WGS. Figure 9.1C and D displays the degree of overlap between samples taken inside and outside Africa. Within Africa, some populations show mixtures of genotypes, primarily those in Senegal and the east coast of Kenya. These are almost certainly due to recent introductions of Aaa back into Africa (Brown et al. 2014; Kotsakiozi et al. 2018b).

Outside Africa, almost no admixture with Aaf is detected, except in Argentina (figure 9.1A). An attempt to eradicate *aegypti* from the New World was carried out between about 1945 and 1970, and it was partly successful. Northern Argentina was an exception; its populations were not eradicated (figure 1.8). Given this unique signature of African ancestry in mosquitoes in Argentina, it is conceivable that these populations represent relicts of the original introduction, a hypothesis being actively explored at this time.

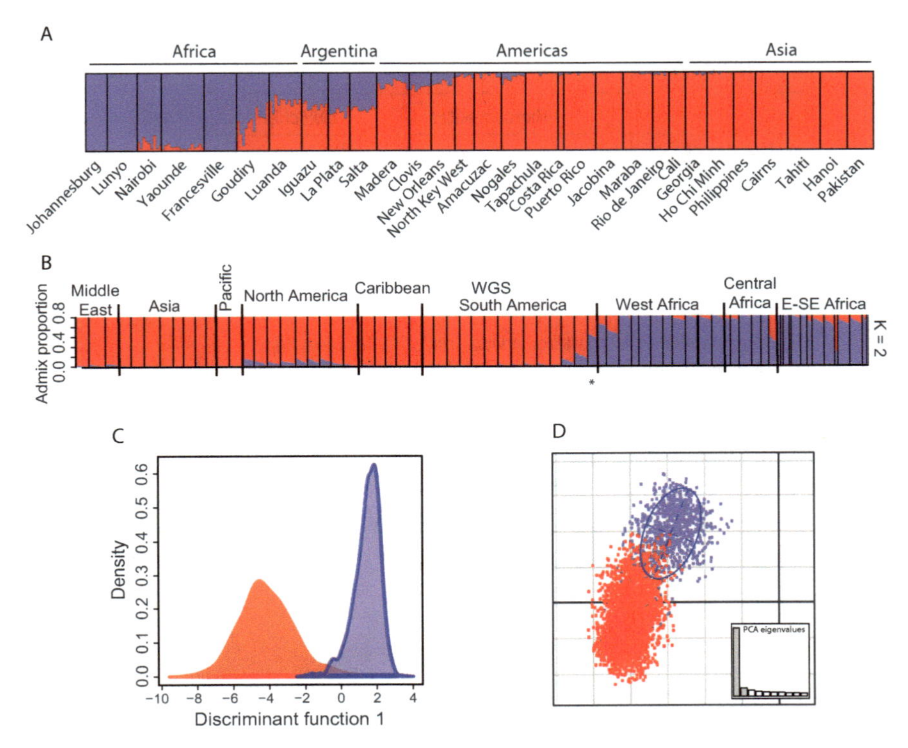

Figure 9.1 Two types of genetic data for *Aedes aegypti* worldwide. In all plots, samples from Africa are in blue, and those outside Africa are in red. Different numbers of population samples are used in each analysis. **A** and **B** are Bayesian STRUCTURE plots (K = 2) for, respectively, 19,000 SNPs from a chip (**A**) and whole genome sequencing (**B**). Both data sets clearly distinguish populations from Africa (largely Aaf) and those outside Africa (Aaa), as well as detecting the same mixed populations in Africa (Senegal and the east coast of Kenya) and outside Africa (Argentina, indicated by an asterisk below the plot in **B**). **C** and **D** are based on 12 microsatellites, with **C** being a discriminant analysis of principal components (DAPC), and **D** showing the first two principal components. SNP chip data are unpublished data from the Powell and Gloria-Soria labs, analyzed by D. Balcazar; WGS data are from Crawford et al. (2024); microsatellite data are from Gloria-Soria et al. (2016a).

Macrogeographic Patterns

Here I consider the differentiation within the two major genetic groups: within Africa and outside Africa. Figure 9.2 shows microsatellite data for New World populations. More detail can be detected when subsets of samples are considered and investigators vary K (the hypothesized

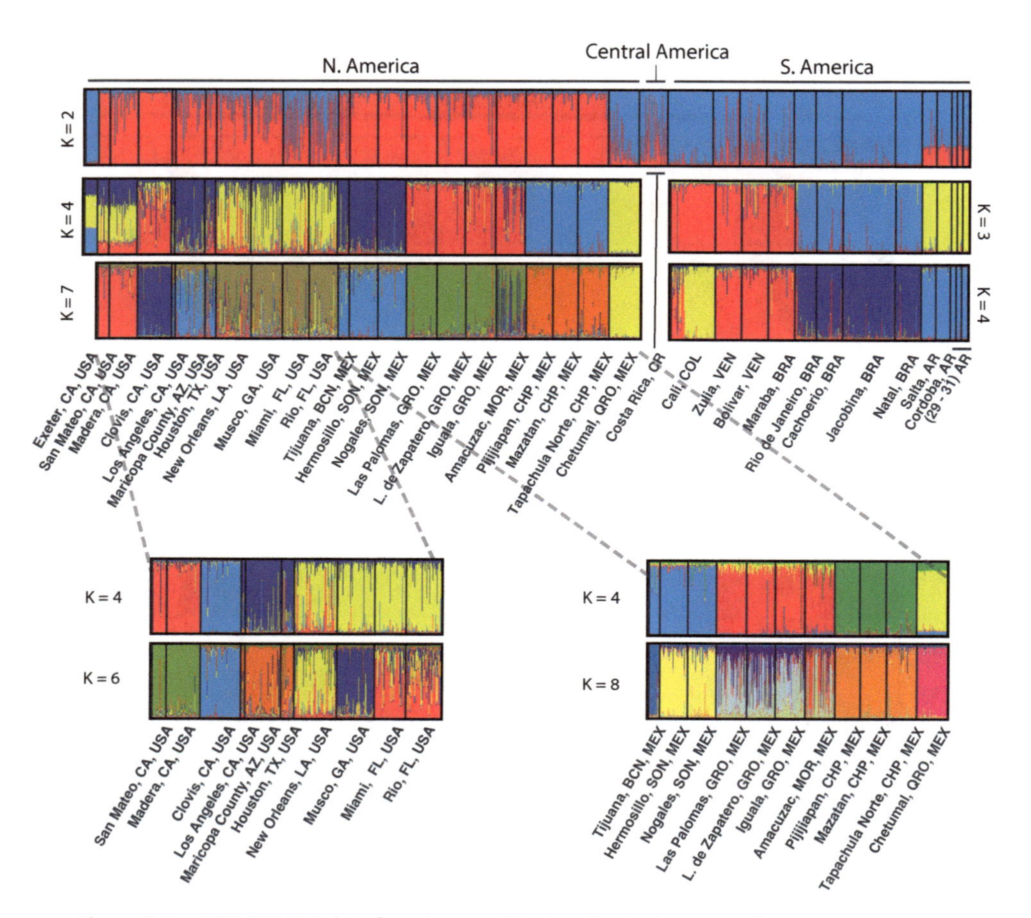

Figure 9.2 STRUCTURE plots for microsatellite data for *Aedes aegypti* from across its distribution. Considerable differentiation is clear, and when the data are subdivided into smaller regions, further local dissimilarities are detected, almost to the point where each population (collection) is unique. From Gloria-Soria et al. (2016a). Similar data are available from the same publication for Caribbean, Asian, and South Pacific populations.

number of units). For the most part, North and South American populations are distinct, as are populations within each of these subdivisions. Gloria-Soria et al. (2016a) provide similar data for Asia and the Caribbean. Ultimately, from these kinds of analyses each population sample can be shown to be genetically unique or to belong to a small subset of geographically restricted sets of populations.

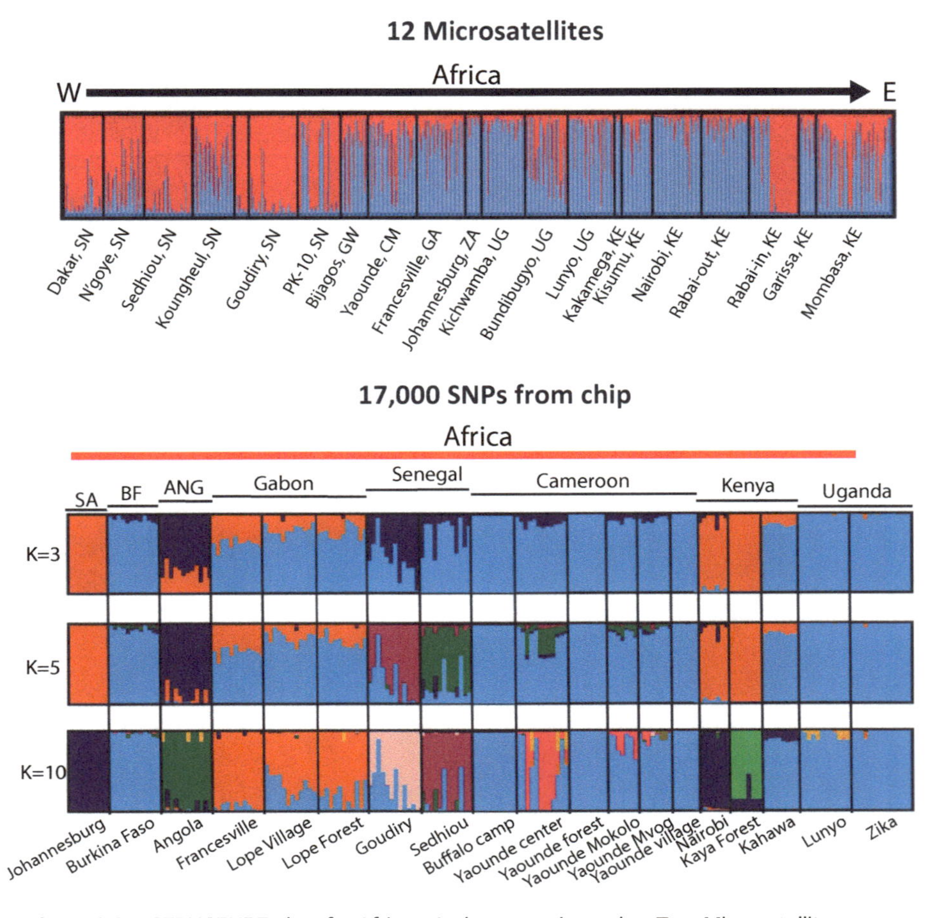

Figure 9.3 STRUCTURE plots for African *Aedes aegypti* samples. **Top**. Microsatellites reveal minimal structuring. From Gloria-Soria et al. (2016a). **Bottom**. Plots exhibit considerably better resolution when using SNP chip data. From Kotsakiozi et al. (2018b).

African Aaf populations are generally less cleanly differentiated with microsatellites, providing only minimal information. Kotsakiozi et al. (2018a) used SNP chip data and did detect significant differentiation within Africa. Figure 9.3 compares the resolving power of these two sources of data (12 microsatellites versus 17,000 SNPs) for African populations. Clearly, the SNP dataset provides more resolution.

It is of some interest to ask why African populations are less genetically differentiated than those outside Africa. African populations are older and, for most of their history, have existed in an environment that had quasicontinuous favorable habitats. This allows more time for migration to genetically homogenize populations across these habitats. Outside Africa, populations are younger, likely have been founded by only a few individuals (founder effects), and exist mostly in human habitats that are discontinuous, thus reducing migration.

Smaller Scale Patterns

As noted above, a partially successful major eradication program to eliminate invasive *aegypti* in the New World was undertaken shortly after World War II (figure 1.8). Brazil was declared free of *aegypti* in about 1955 but was reinvaded by 1970. Contemporary Brazilian populations form two major clusters: a northern and a southern one (Monteiro et al. 2014; Kotsakiozi et al. 2017b). The northern one was likely founded from Venezuela, which was never free of *aegypti* during the eradication program. ABC analysis was consistent with northern populations breaking out from Venezuela in about 1970 (Kotsakiozi et al. 2017b). The origin of the southern cluster is less clear, possibly from the Caribbean or derived from the northern one. Maitra et al. (2019) also detected two major clusters of *aegypti* in Brazil, but their data do not precisely coincide with that of Kotsakiozi et al. (2017b).

Da Costa–Ribeiro et al. (2016) used microsatellites to study 14 districts in Rio de Janeiro over two years. These districts were genetically distinct, even though some were as close as 1 km and others were a maximum of 30 km apart. More intriguing was the observation that the degree of genetic distinctness varied over seasons, with significantly more differentiation in the wet season. This is likely due to more larval breeding sites, so females do not need to move far to find a suitable place to lay eggs (chapter 3). Another study incorporating a temporal perspective was carried out in Thailand (Olanratmanee et al. 2013). They found fluctuating levels of genetic differentiation in three out of four samples taken over a year. This was attributed to having only a few particularly productive breeding containers contributing the bulk of the adults. Six populations in Haiti have also been studied for microsatellite allele frequencies (Lovin et al. 2009). Ho Chi Minh City has been studied with regard to dengue vector competence and genetic differentiation (Tran Khanh et al. 1999).

Seasonal effects on genetic differentiation among *aegypti* populations have also been examined in Cebu City in the Philippines (Sayson et al. 2015). The Cebu City study was done over time and, contrary to the Rio de Janeiro one, greater genetic differentiation was found in the dry season, compared with the wet season. The authors attributed this to a contraction of populations as breeding sites dried up. Sampling across seasons was also done in Manila, but it was uninformative, as there was no clear genetic structure across the city at any specific time period (Carvajal et al. 2020). The authors conclude that this is likely due to passive transportation in a locality with a high-density human population.

Another region of South America receiving attention is Bolivia (Paupy et al. 2012). Two genetic clusters were detected: one widespread throughout the country, and a smaller cluster confined to rural areas. The latter cluster had genetic affinities to West African mosquitoes, consistent with the notion that *aegypti* may not have been eradicated from this part of South America. Like Argentina, these populations have some genetic signature from the original introduction from Africa.

Islands in the South Pacific have also been the focus of population genetics studies. Calvez et al. (2016) genotyped nine samples, across about 5,000 km, for microsatellites. The islands clustered into four or five reasonably distinct units. Not unexpectedly, Australia and Southeast Asia are implicated as the sources of these populations. More surprising is some indication of possible American contributions, based on mtDNA (with the reservation expressed in note 3 below). *Ae. aegypti* could possibly have been introduced during World War II, when considerable numbers of troops traveled from the United States to these islands. Paupy et al. (2000) studied some of these same islands and found that populations sampled in regions of high human density were more genetically differentiated than those in rural areas. These authors attributed this to consecutive treatments of insecticides in populated regions.

A comparative population genetics study was carried out on *aegypti* and *albopictus* in Southeast Asia and the South Pacific, using a large dataset of 50,000 SNPs (T. Schmidt et al. 2020). This study was the first to explicitly use methods from landscape genetics to explain genetic patterns (see below for further studies). It considered geographic distance and both shipping and aerial transport networks. Isolation by distance (IBD) was found, but no significant correlation could be detected relating the level of transport to genetic differentiation. Of note, genetic differentiation of *albopictus* in this same area displayed the opposite result: no significant IBD was detected, but there was significant genetic differentiation associated with both aerial and marine traffic. More details on Australian and Vietnamese populations are in Endersby et al. (2009).

The contrasting genetic differentiation of *aegypti* in Florida and California in the United States is striking. Over the same geographic space, Florida populations are remarkably homogeneous, while southern California populations are highly structured (Pless et al. 2020). Florida likely has some of the oldest populations outside Africa, as this was a site of early introductions and, as far as is known, has continuously had *aegypti* since this species was introduced; it survived the eradication program in the mid-20th century (figure 1.8). California has "officially" had *aegypti* since 2013, but likely it was introduced 30 or so years earlier (see the next section). In addition to age, contrasting environments are likely also involved. Southern California has an annual dry season of six to eight months, whereas Florida has high levels of moisture year round.

Timing Events

Placing time estimates on when populations separated is also possible with population genetics.[4] An example for *aegypti* is determining when it was introduced into California. Despite active surveillance in the state, until 2013 this species had not been reported, while, since then, it has been consistently found. In an attempt to identify the origin of the introduction, Gloria-Soria et al. (2014) used microsatellites, and Pless et al. (2017) used 16,000 SNPs. A timing analysis is illustrated in figure 9.4A, where two things are apparent. First, California was invaded

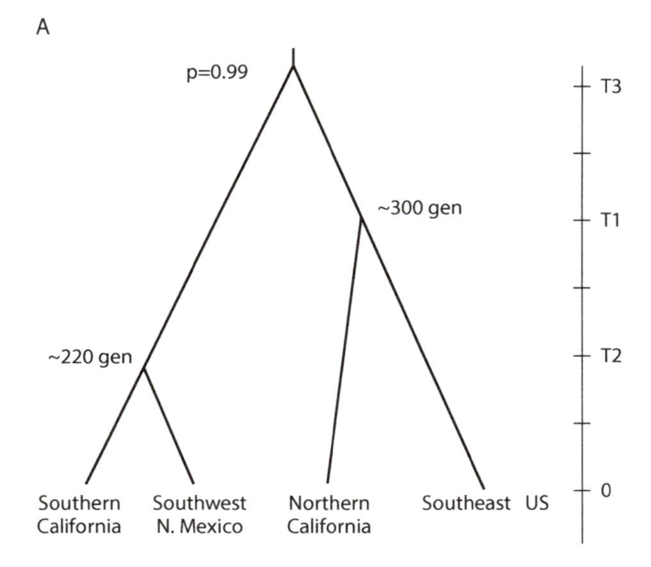

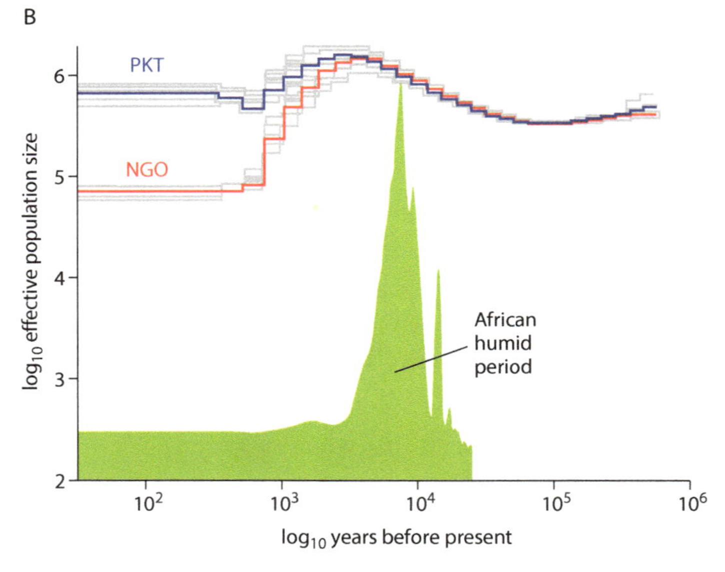

Figure 9.4 **A**. Approximate Bayesian computation (ABC), providing time estimates for the introduction of *Aedes aegypti* into California. The p value indicates strong support for this branching pattern, compared with alternatives. The time estimates for the two nodes are in generations. From Pless et al. (2017). **B**. Timing of effective population sizes (**top**), using cross-coalescence for two West African populations of *Aedes aegypti*: one primarily a human-biter (PKT), and the other preferring nonhuman animals (NGO). This supports the hypothesis that the recent drying of West Africa led to a divergence of population host choices for blood meals. The bar graph (**below**) shows inferred rainfall data. From Rose et al. (2023).

from two sources, one going to northern California and the other to the southern part of the state. Most striking are the times associated with these introductions. Using 10 generations per year (although this is likely an overestimate, given the cool winter temperatures in the Central Valley of California, where *aegypti* is breeding), the invasion must have happened at least 20–30 years ago, yet was detected only one year before these sample were taken. This implies that *aegypti* can be cryptic for some time before being reported.[5] A similar story holds for Black Sea populations (see below).

Rose et al. (2023) used a novel, sophisticated cross-coalescence method and WGS data to generate time estimates of events. Figure 9.4B shows the results. The mainly human-biting populations of African *aegypti* broke off from primarily animal-biting ancestral types shortly after a dry period set in. This is consistent with what chapter 1 described as the recent history of *aegypti*, based on the then-available genetic data, epidemiology, and collection information. Support from so many sources increases confidence in the conclusions that were reached. This figure also demonstrates that WGS data can be used to estimate demographic parameters, like effective population size (figure 9.4B, y-axis).

Finding the Missing Link in Museum Specimens

In describing the history of *aegypti*, strong evidence was presented for Asian populations being derived from the New World, but no solid evidence was presented with regard to the route taken. Powell et al. (2018) speculated that ships returning from the New World to ports in the Mediterranean brought back *aegypti*, and that these now-extinct populations were the source for Asia. Today, Mediterranean *aegypti* can be found in museums. Prior to analyzing these specimens, an extensive SNP chip tree was built, based on about 19,000 SNPs, which included samples from around the Black Sea, a region thought to have been free of *aegypti* for many decades (figure 9.5; also see Kotsakiozi et al. 2018b). Black Sea populations are genetically intermediate between those in the New World and Asia, but it was not clear whether these were recent invasions from Asia or possibly a remnant from the originators of Asian populations. Sufficient DNA was obtained from the Mediterranean

Asia
Black Sea
South America
Caribbean
North America
Africa
(A) Vietnam2
(A) Vietnam2
(A) Thailand
(A) Thailand
(A) Vietnam I
(A) Vietnam I
(A) Philippines
(A) Philippines
(A) SaudiArabia
(A) SaudiArabia
(A) Pakistan
(A) Pakistan
(A) Australia
(A) Australia
(A) Tahiti
(A) Tahiti
(B) Georgia
(B) Georgia
(B) Turkey
(B) Turkey
Eight Mediterranean museum samples
(D) Jacobina, Brazil
(D) Jacobina, Brazil
(D) Rio, Brazil
(D) Rio, Brazil
(D) Maraba, Brazil
(D) Maraba, Brazil
(D) Cali, Colombia
(D) Cali, Colombia
(C) PuertoRico
(C) PuertoRico
(C) Trinidad
(C) Trinidad
(E) KeyWest, Florida
(E) KeyWest, Florida
(E) CostaRica
(E) CostaRica
(E) Tapachula, Mexico
(E) Tapachula, Mexico
(E) Clovis, California
(E) Clovis, California
(E) Madera, California
(E) Madera, California
(E) Amacuzac, Mexico
(E) Amacuzac, Mexico
(E) Tijuana, Mexico
(E) Tijuana, Mexico
(E) Nogales, Mexico
(E) Nogales, Mexico
(E) Georgia, USA
(E) Georgia, USA
(E) NewOrleans, Louisiana
(E) NewOrleans, Louisiana
(F) Senegal1
(F) Senegal1
(F) Senegal2
(F) Senegal2
(F) SouthAfrica
(F) SouthAfrica
(F) Kenya
(F) Kenya
Ae. ae. formosus
mascarensis

museum specimens to obtain DNA sequences. Figure 9.5 shows where these Mediterranean samples fit into the larger picture of *aegypti* phylogeny. Extinct Mediterranean populations were genetically nearly identical to present-day Black Sea populations, providing strong support that the Mediterranean was a stepping stone between the New World and Asia. Placing a time estimate on the Black Sea samples suggested that they had continuously been in existence for 100–150 years and were not extinct, as had been reported. This is a second example of *aegypti* populations being cryptic for some time before being erroneously reported as *re*introductions.[5]

Genetic Stability of *Aedes aegypti* Populations

Obviously the patterns and inferences drawn from population genetics studies strictly apply to the time when populations were sampled. How stable are these patterns? Gloria-Soria et al. (2016b) examined 14 locales where two or three collections were taken, one to three years apart. While some temporal fluctuations were detected, they are not great and did not change the patterns of geographic genetic differentiation in any significant manner. The data allow tests to detect bottlenecks between collections, which were found in a minority of localities. Rasic et al. (2015) used a variety of genetic markers in studies in Yogyakara, Indonesia, and found similar patterns. While there were minor fluctuations from year to year allowing an estimate of N_e (see later in this chapter), these temporal fluctuations did not significantly change the geographic pattern of divergence among populations.

Other indications of an even longer temporal genetic stability for *aegypti* populations come from comparisons of patterns between early investigations in the 1970s and recent studies since 2000. One example

Figure 9.5 (*opposite*) Maximum likelihood tree for *Aedes aegypti* collections from around the world, based on ~19,000 SNPs from a chip (Kotsakiozi et al., 2018a). Eight museum specimens of now-extinct Mediterranean samples from southern Europe and northern Africa have also been sequenced (unpublished data from the author's laboratory). These specimens are virtually identical to the samples from Turkey and Georgia that, in this phylogenetic tree, are intermediate between the New World and Asian populations. This supports the hypothesis that the Mediterranean was a stepping stone in the foundation of Asian populations from the New World.

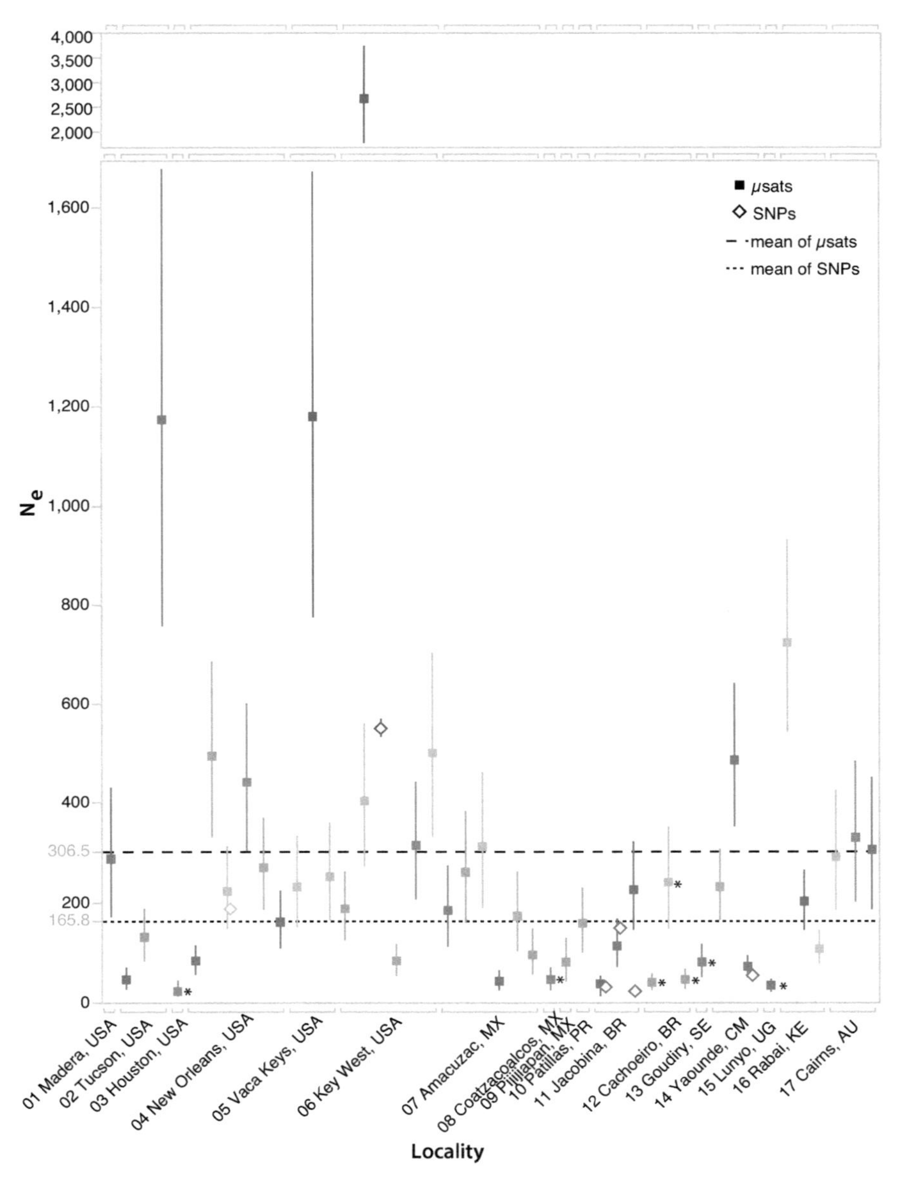

Figure 9.6 Effective population size (N_e) estimates for several *Aedes aegypti* populations, inferred from temporal sampling. Points with larger error bars are for microsatellites, while the smaller error bars represent data from the SNP chip. From Saarman et al. (2017).

of this stability is the genetic discontinuity between the southeastern and southwestern United States. The 1970s work detected a genetic break between Beaumont and Houston, cities about 100 km apart in Texas (Wallis et al.1983). Gloria-Soria (2016a) and Pless et al. (2022) found this same break in collections taken some 35 years later (figure 9.6). Note that in the more-recent work, the position of this break is less well defined, as *aegypti* had disappeared between New Orleans and Houston, likely due to the invasion of *albopictus*.

Effective Population Sizes

Effective population size (N_e) is a theoretical parameter that measures the strength of genetic drift. This would be equal to the census size (N_c) if every individual reproduced and left the same number of offspring. In reality, N_e is less than N_c, due primarily to unequal contributions from individuals and unequal numbers of males and females. One of the more accurate ways to measure N_e in field populations is by taking genotype samples at the same location but at different times and then estimating the number of generations between samples. An N_e that would account for observed changes in allele frequencies is then estimated, assuming genetic drift is the only factor involved.

Figure 9.6 shows the results for several populations where more than one sample is available (Saarman et al. 2017). While the mean N_e is 166 (from SNP data) or 307 (from microsatellite data), N_e less than 100 is not uncommon. Other studies in Southeast Asia and Australia, using different kinds of markers, have reached essentially the same conclusions. N_e estimates between 166 and 692 were found at three localities (table 9.2).

Table 9.2 Estimates of N_e in *aegypti* populations based on temporal sampling

Locality	Data	N_e estimate	Reference
Australia	microsatellites	692	Endersby et al. (2011)
Indonesia	microsatellites and SNPs	467	Rasic et al. (2015)
Thailand	microsatellites and SNPs	166	Olanratmanee et al. (2013)

Note: Observe the consistency of this table's data with the data in figure 9.6.

Census size estimates (N_c) have also been made for *aegypti* using mark-recapture methods (table 7.1). Most studies estimate populations of several hundred to a few thousand. Thus the ratio of N_e to N_c in *aegypti* is 10%–30%, in line with most natural populations of animals.

Landscape Genetics

A relatively new approach to identifying environmental factors affecting the level of movement among populations (migration rates, or *m* in population genetics) is called "landscape genetics." Several features of landscapes where populations have been sampled are quantitatively assessed, often with the aid of global satellite imaging data.

Pless et al. (2021) carried out such a study along the southern tier of the United States and parts of Mexico and the Caribbean (figure 9.7). The genetic datasets were microsatellite allele frequencies for collections from 38 sites. Their environmental dataset contained 29 variables, such as climate, land cover, human infrastructure, and so forth. A novel statistical approach—machine learning—was used, and the correlation (R) between the model's predicted genetic distance and the actual distance was a remarkable 0.83. Environmental variables contributing the most to the accurate predictions were maximum temperatures and slope (likely related to the retention of moisture), followed by barren land cover and human density. Of note, environmental factors identified as providing particularly suitable habitat for *aegypti* (Dickens et al. 2018) are not the ones identified here as promoting migration. This implies that *aegypti* in highly suitable environments have less incentive to move.

Because there is good evidence that this range in the southern part of the United States was colonized from east to west, models of serial founder effects were also tested—that is, each step from east to west being accomplished each time by only a few individuals. Pless et al. (2022) found that this model works quite well. Allelic richness decreases from the East Coast of the United States to California. Northern and southern California populations differ in allelic richness, likely reflecting two introductions (figure 9.4).

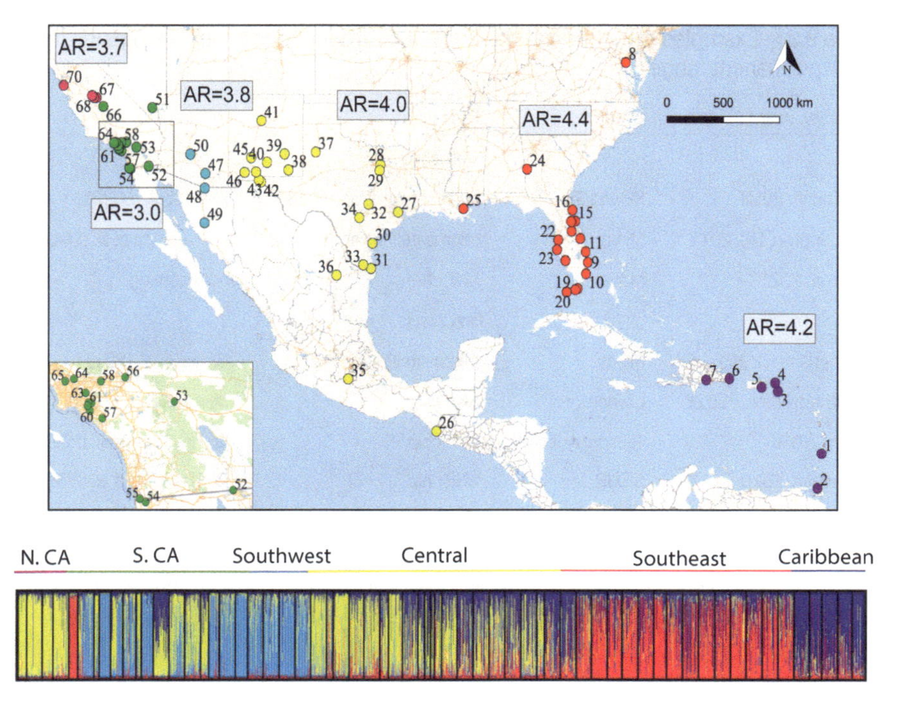

Figure 9.7 Data used in landscape genetics studies of *Aedes aegypti*. **Top**. Data using microsatellites. Sparse sampling along the northern coast of the Gulf of Mexico is due to a lack of populations when the study was done, likely due to an invasion of *albopictus*. From Pless et al. (2021). **Bottom**. Much of the same data was used in detecting serial founder events, as indicated by allelic richness (AR). Some ambiguity in the California samples probably reflects two independent introductions into California. From Pless et al. (2022).

Identifying Origin of New Introductions

We can delineate a stable year-round breeding distribution of *aegypti* (figure 1.6), but because it is an invasive species closely tied to human habitats (outside Africa), it regularly makes an appearance outside its stable range. Samples of new introductions have been genotyped to determine their origins. This has only recently become possible, due to the accumulation of a large population genetics database for the species (table 9.1). Table 9.3 summarizes eight such studies. In most cases the identified locale of origin makes sense. For example, the Netherlands is a center for recycling used tires. Unmounted automobile tires are a

Table 9.3 Examples of identifying the origin for introductions of *aegypti* outside its usual permanent range

Locality, year	Origin	Data used	Reference
		Established	
California, 2013	SE and SW US	microsatellites and SNPs	Pless et al. (2020)
Washington, DC, 2011	Miami	microsatellites and mtDNA	Gloria-Soria et al. (2018b)
Cyprus, 2021	Madeira	SNP chip	forthcoming
		Temporary	
Netherlands, 2010	Miami	microsatellites	Brown et al. (2011)
Southern France, 2018	Cameroon	SNP chip	Jeannin et al. (2022)
Utah, 2019	Tucson, Arizona	SNP chip	Gloria-Soria et al. (2022a)
Nebraska, 2019	SE US	SNP chip	Gloria-Soria et al. (2022a)
Las Vegas, 2017	S. California	microsatellites	Pless and Raman (2018)

Note: In the upper three rows, populations persisted for multiple years. In the lower five rows, as far as is known, the introductions quickly disappeared.

common mode of passive transport for container-breeding mosquitoes (Reiter and Sprenger 1987). Similarly, Marseille is a port with regular shipping traffic from West Africa (Jeannin et al. 2022).

Lab Strains

Because *aegypti* is frequently used in laboratory studies, a few established strains have become standards in lab work. It is of some interest to determine (a) if these strains are genetically similar to their purported site of origin, (b) whether multiple copies of the strains in different labs are genetically identical (or at least largely similar), and (c) whether genotype signatures can be developed to ensure that these standard strains are the same in future uses. The ROCK (Rockefeller) strain, generally thought to have come from Cuba in 1961, is the most widely used lab strain (Kuno 2010). Genetic analyses of three copies of ROCK from different laboratories found two to be genetically uniform, but a third was quite distinct (Gloria-Soria et al. 2019). More intriguing was the Liverpool strain, which was used in deriving the original complete genome assemblies and was thought to have originated in West Africa.

Three out of four copies of this strain from various labs were most similar to Asian populations, while the fourth copy was genetically distinct, being closer to the Florida (Orlando) strain. A commonly used cell culture line of *aegypti* (AAG2) was derived from the ROCK strain, and multiple contemporary copies of this cell line remain genetically almost identical to most ROCK strains today.

Anopheles gambiae Complex

From a population genetics viewpoint, the *gambiae* complex differs from the other two mosquitoes in having a narrower distribution and being replete with chromosomal inversion polymorphisms. The former makes population genetics easier, while the latter complicates interpretations. Paracentric inversions suppress recombination between the breakpoints, meaning that the inverted segment is inherited intact, like a single Mendelian gene (chapter 8). (Double crossovers and gene conversion are events that negate this rule.) Mutations giving rise to inversions—two breaks in a chromosome arm that then get repaired with the fragment having the opposite orientation—are very rare, and it is generally thought that the new inverted chromosome arose only once. This means that all the alleles in a newly inverted chromosome represent a single haploid, which will be inherited intact.

I will first present evidence that at least some of the inversions in the *gambiae* complex are adaptations to environmental heterogeneity—in particular, aridity and larval breeding sites. Then I will discuss some studies using molecular approaches, including (ultimately) WGS. Chapter 10 (on speciation) will consider inter-taxa patterns.

Inversions

Figure 2.4 illustrates the high quality of polytene chromosome preparations from *gambiae*, and figure 8.2 shows the extent of polymorphic inversions. While this looks complex, there are only six common inversions in 2R, and one in 2L (shown by brackets below the chromosomes in figure 8.2). These are the ones that play a large role in *gambiae* biology.

Selection acts to maintain the frequencies of inversions in the *gam-*

biae complex, demonstrated by gene arrangements in the common inversions in both arms of the second chromosome. Figure 9.8A shows the frequencies of chromosome 2 arrangements in the left (2L) and right (2R) arms. The important point is that populations sampled from the left of this figure come from very humid environments on the Niger River delta, progressing to drier climates when moving to the right, and ending at the edge of the Sahara Desert. The implications are that arrangements 2La, 2Rd, and 2Rbc are adapted to arid environments, while 2La$^+$, 2Rd$^+$, and 2Lb$^+$ have higher fitness in humid conditions.[6] While clines of this sort implicate a role for natural selection acting on the variants, it is possible that such patterns could arise by chance.[7] This interpretation became much less tenable, however, with the observation of temporal seasonal shifts in frequencies of these same gene arrangements. The arrangements following the various levels of aridity geographically track the wet and dry seasons temporally (figure 9.8B). A final piece of evidence for the association of these inversions with aridity/humidity comes from small-scale distributions. In West Africa at night, humidity is lower inside huts than outside. Thus the inversions implicated in being associated with lower humidity (2Rd and 2La) are more frequent in indoor collections than in outdoor ones taken less than 100 meters away (Coluzzi et al. 1979; Powell et al. 1999).

An added complexity in interpreting older data (like that in figure 9.8A) is the recent understanding of cryptic taxa in the *gambiae* complex. The collections analyzed may well be a mixture of taxa that are recognized today. But this is irrelevant in concluding that these inversions are involved in adaptations to mesic/arid habitats. As discussed in chapter 10, there is good evidence that there has been introgression between taxa, so they share monophyletic inversions that confer similar phenotypes (adaptations to varying levels of humidity). In their temporal study, Touré et al. (1998a) recognized the M and S molecular forms and confirmed the association of these inversions with levels of aridity.

Chromosome 2 inversions were studied in more detail in Cameroon over a similar arid-to-moist gradient (Cheng et al. 2012). Their samples more reliably came from one taxon, *gambiae* s.s., and their data were

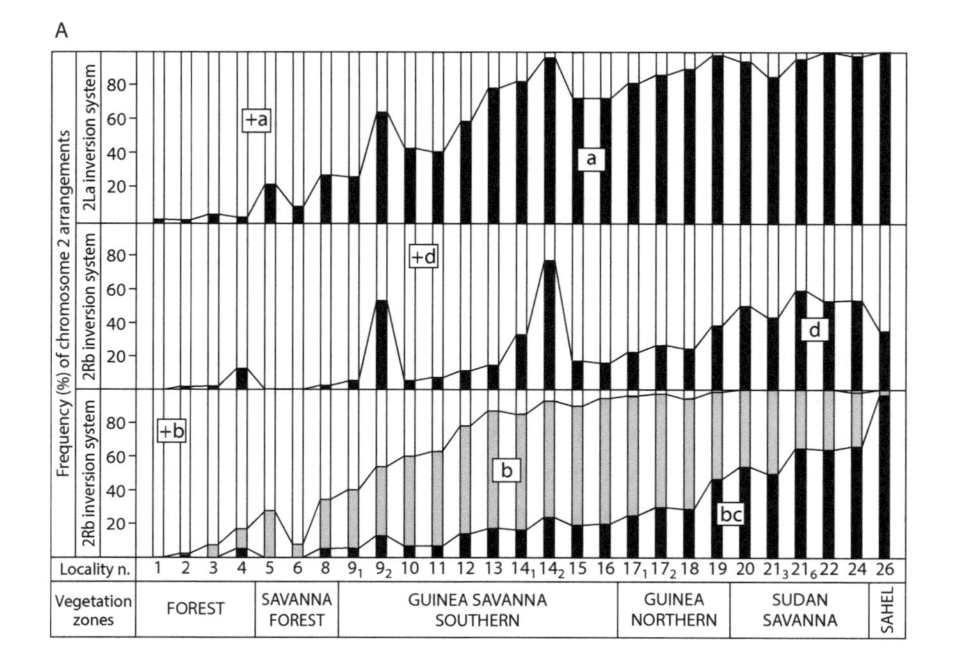

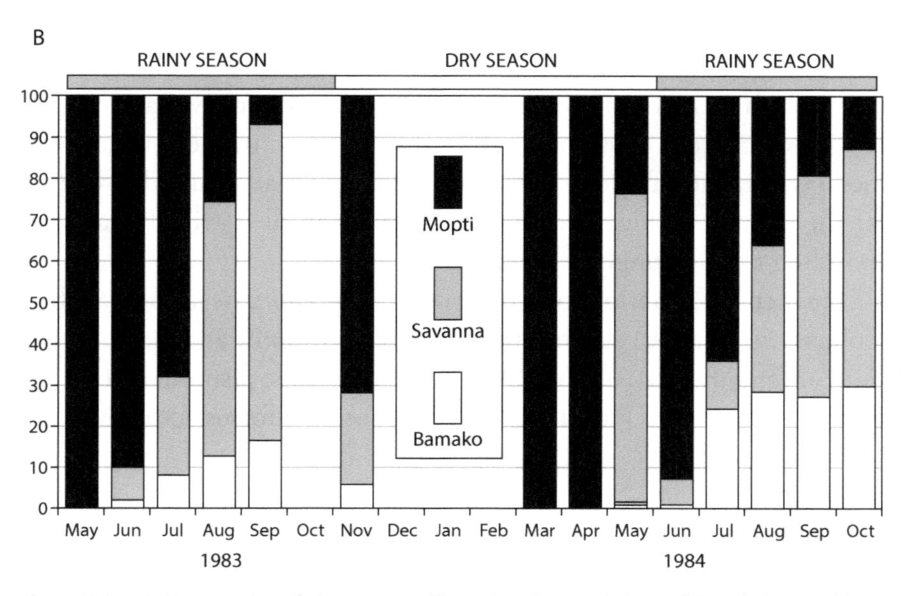

Figure 9.8 **A**. Frequencies of chromosomal inversions in populations of *Anopheles gambiae*. The gradient (left to right) is from western coastal Nigeria north to the southern Niger Sahel. From Coluzzi et al. (1985). **B**. Seasonal abundance of the chromosomal forms in indoor-resting *gambiae* s.s. females in Banambani, Mali, in different months of two consecutive years (1983–1984). No collections were carried out in October and December 1983, and in January and February 1984. Sample sizes ranged from 48 to 525 specimens. Note the chromosomal type whose geographic distribution (in **A**) would indicate that it is the most dry-adapted form predominates in the dry season whereas the wet-adapted inversions increase through the rainy season. From Touré et al. (1998a).

from DNA sequences along the entire genome. They found that panmixia (random mating) prevailed, with no indication of reproductive isolation, as would be expected if taxa were mixed. The prevalence of 2La and 2Rb in drier regions was confirmed. Differentiation along the genome was remarkably even, except for the regions covered by 2La and 2Rb, which had considerably elevated divergences. This implies sufficient gene flow among populations to homogenize gene frequencies except where countered by selection. Most telling was the observation that gene exchange between the inverted and noninverted chromosomes did occur, especially away from the breakpoints, meaning that the sequences between breakpoints are not completely isolated. The authors attributed this to rare double recombinants and/or gene conversions.

Cuticular hydrocarbons are known to be associated with desiccation resistance in many insects. Both cuticular hydrocarbons and the thickness of the cuticle have been examined in *gambiae* collected at locales with different levels of aridity, with equivocal results (Reidenbach et al. 2014). Cheng et al. (2018) concluded that overall differences in energy metabolism were a more likely possibility than cuticle variation. This finding came from a systems analysis that included transcriptional profiles, gender, and humidity regimes. Environment and humidity had by far the greatest influence on transcriptional profiles, with females responding more strongly than males.

Further information on inversions in *gambiae* is also available. Coluzzi et al. (2002) summarized the distribution of about 200 inversions among the various taxa. Della Torre et al. (2005) summarized considerable data on frequencies of the chromosomal forms across Africa.

Microsatellites

In a series of studies across much of Africa, Lehmann and colleagues document a consistent pattern of genetic variation, based on microsatellites (Kamau et al. 1998; Lehmann et al. 1997, 1999, 2003). Two widespread genetic subdivisions were detected: a northwestern one (in Senegal, Cameroon, Nigeria, Gabon, Democratic Republic of the Congo, and western Kenya) and a southeastern one (in eastern Kenya, Tanzania, Malawi, and Zambia). The level of genetic differentiation within each of these two groups is very low, with F_{ST} <0.02, even between populations separated by 6,000 km (e.g., western Kenya and Senegal, with F_{ST} = 0.016). In contrast, the differentiation between populations from different major subdivisions have an average F_{ST} >0.20. The Rift Valley runs between these groups.

Onyabe and Conn (2001a) similarly found very little genetic differentiation in microsatellites from arid northern Nigeria and the more humid forest in the south when they excluded loci on chromosome 2. This supports the observation that the strong inversion gradient in figure 9.8 is due to selection. Migration among populations is high enough to homogenize the frequencies of neutral genetic variants (as is generally assumed for microsatellites).

Using microsatellites, Donnelly et al. (2001) inferred that Kenyan populations west of the Rift Valley had undergone a recent population expansion, while those east of the valley had undergone both a recent expansion and a bottleneck. This could explain the relatively high linkage disequilibrium found in a Kenyan sample (see figure 2.2).

Landscape Genetics

A formal landscape genetics study was performed across Kenya using microsatellites (Hemming-Schroeder et al. 2020). They found that connectivity for *gambiae* was affected most by tree cover, and by cropland for *arabiensis*. Of note, a landscape genetics analysis of the *Plasmodium* malaria parasite was performed in the same region (Hubbard et al. 2023). In contrast to the mosquito vector, these authors found relatively little genetic differentiation in the parasite, although Lake Victoria and elevation had minor effects on gene flow among parasite populations.

SNP Chips

While at least two SNP chips have been developed for the *gambiae* complex, they have been used sparingly for population genetics. Neafsey et al. (2010) assayed the variation across several populations of *gambiae*, *coluzzii*, and *arabiensis* for 400,000 SNPs on their chip. Their main interests were to (a) identify regions of the genome differentiating the species and (b) identify selective sweeps.[8] Intriguingly, the most differentiated regions of the genome between the species were the same as those implicated in selective sweeps.

Nwakanma et al. (2013) used this SNP chip to study differentiation between the then-recognized molecular forms M and S (discussed further in chapter 10).

Insights from WGS

The first completely sequenced genomes of mosquitoes were for the *gambiae* complex (Holt et al. 2002). Since then, literally thousands more individuals have been sequenced, spanning the entire *gambiae* complex. Landmark studies have been done by the *Anopheles gambiae* 1000 Genomes Consortium (2017, 2020). Figure 9.9 summarizes patterns of differentiation. Collections generally cluster by country, especially in the eastern and western parts of Africa. In the middle part of the continent, however, there is more mixing, especially in Cameroon. These two papers from the consortium also present further analyses, including levels of genetic diversity (1%–2% heterozygosity) across the genomes, estimates of migration rates, and changes in population size over time. A further study of these data focused on detecting selective sweeps (Xue et al. 2020).

H. Schmidt et al. (2019) added 111 WGS samples from the Comoros Islands off the east coast of Africa. While their sample sizes are small (both in regard to localities and individuals), they obtained modest evidence of serial founder effects proceeding from west to east.

Anopheles arabiensis

This species' distribution largely overlaps that of *gambiae* s.s. (figure 1.9), although it extends farther north and south into drier regions. While there is clear evidence that there has been introgression between *arabiensis* and *gambiae* s.s. since their initial split (chapter 10), this has occurred long enough ago that, for population genetics analyses, contemporary populations are not compromised. Plus, *arabiensis* has long been recognized as a distinct species, so older literature had kept it separate from *gambiae* s.s.

An. arabiensis was included in the studies of Hemming-Schroeder

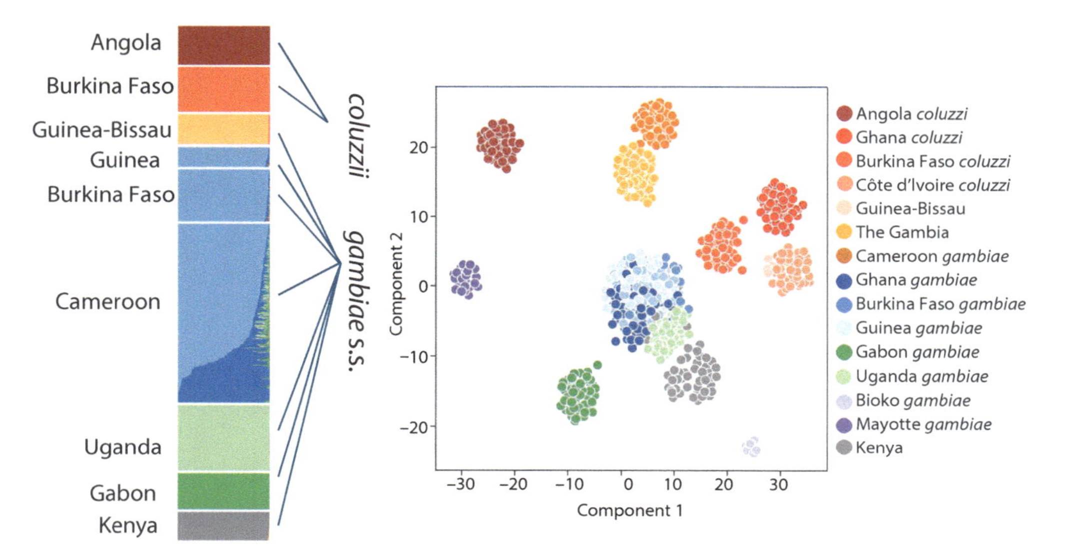

Figure 9.9 Population structure of two members of the *Anopheles gambiae* complex: *gambiae* s.s. and *coluzzii*. **Left**. A STRUCTURE plot for WGS data. From *Anopheles gambiae* 1000 Genomes Consortium (2017). **Right**. A UMAP plot, similar to PCA, for some of the same populations plus others, using biallelic SNPs from chromosome 3. Equal sample sizes were used by randomly sampling 50 individuals from each locality. From *Anopheles gambiae* 1000 Genomes Consortium (2020).

et al. (2020) and Donnelly et al. (2001) discussed above. Like *gambiae*, the Rift Valley is a significant barrier to gene flow in *arabiensis*, and this species also underwent a recent population expansion. More telling are studies that have focused exclusively on *arabiensis*. Onyabe and Conn (2001b) studied a similar north (dry) to south (wet) transect through Nigeria, as had been done previously for *gambiae* (figure 9.8). They found decreasing allelic diversity from north to south, suggesting recent serial founder effects. Fully 78 alleles found in the northern savanna populations were absent in the moist south, while only a single allele found in the south was missing in the north.

Results from other *arabiensis* studies seem contradictory. In some cases, very little genetic differentiation was found, such as in Muturi et al. (2010), who were working in Kenya, and Nyanjom et al. (2003), working in Ethiopia and Eritrea. In other studies, distinct genetic subunits were detected. For example, in Sudan, Mustafa et al. (2021) documented F_{ST}s up to 0.24. Ng'habi et al. (2011) found marginally more genetic differentiation among populations of *arabiensis* than for *gambiae* in southern Tanzania.

So few population genetics studies have considered other members of this complex (*merus*, *melas*, *quadriannulatus*, and *bwambae*) that these species will not be discussed.

Culex pipiens

It is difficult to synthesize and summarize what is known about the population genetics of this complex. One issue is the inherent complexity of multiple taxa and the numerous ways taxon designations have been used in different studies. A second issue is that the data are very heterogeneous, coming from many different methods, as well as varying considerably in the number of genes used and their samples sizes. Nevertheless, in the following I have tried to reach a synthesis.

Single Taxon Studies

Wilke et al. (2014) is one of the few studies to include a single species, *quinquefasciatus*, across Brazil. Figure 9.10 summarizes their results. Moderate genetic structure is apparent, with K = 2 for northern and southern populations displaying the greatest differentiation. When each population is considered (K = 10), some populations seem quite distinct (e.g. COL, LPL, and TER), but most show a considerable sharing of genotypes.

Another study of *quinquefasciatus* (on a much smaller scale) was conducted on the "Big Island" of Hawaii (Keyghobadi et al. 2006). Some degree of genetic structuring was detected, and, although particular attention was paid to the possible effects of elevation, these were not found. This mosquito is of particular importance in Hawaii, as it transmits avian malaria, which has caused some endangered birds to become extinct and threatens many others.

Multiple-Taxa Studies

Most population genetics studies on the *pipiens* complex have included more than one taxon. Examples from Asia, Europe, and North America will be discussed here.

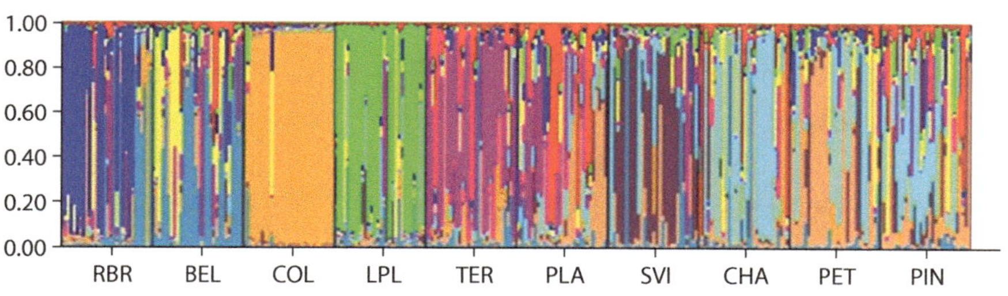

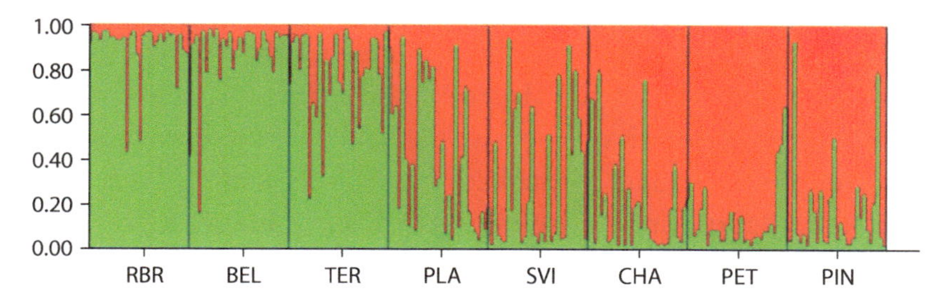

Figure 9.10 Top. Map showing *Culex quinquefasciatus* collections across Brazil. **Bottom.** Upper STRUCTURE plot is for K = 10; lower is for K = 2. Data are from six microsatellite loci. From Wilke et al. (2014).

Cui et al. (2007) studied *pallens* and *quinquefasciatus* along a 2,000 km transect in China. They found little evidence of isolation by distance, unless populations were separated by hundreds of kilometers. They also reviewed a number of other population genetics studies of *pipiens* s.l. from several continents and concluded that throughout its distribution, significant genetic differences are detectable only at 500–1,000 km, regardless of whether the taxon is *pipiens* s.s., *quinquefasciatus*, or *pallens*.

Gomes et al. (2009) studied a locality in Portugal where the two forms of *pipiens* (*molestus* and *pipiens* s.s.) are sympatric. One notable finding was that, while individuals generally fell into two distinct genetic groups (representing the different taxa), there was more introgression of *molestus* genes into *pipiens* than vice versa. Yurchenko et al. (2020) produced WGS data on both *molestus* and *pipiens* in two populations each from North America and Eurasia (figure 9.11). A similar conclusion was drawn with regard to more introgression of genes from *molestus* into *pipiens*. This may not be surprising, given the different mating patterns for the two taxa, since *pipiens* requires a swarm and a large space, whereas *molestus* can mate, without swarming, in a confined space (chapter 5). A *molestus* male that emerges from its confined habitat to the aboveground habitat of *pipiens* could mate and leave genes. A *pipiens* male entering the confined underground habitat of *molestus* is much less likely to be successful in this regard. Gomes et al. (2015) studied similar sets of populations and reached similar conclusions.

Kothera et al. (2013) sampled *pipiens* s.l. populations from southern California to northern California, predicting that *quinquefasciatus* should be found south of 36° N latitude, and *pipiens* s.s. above 39° N. A fairly sharp genetic break was observed at 36° N, but there was considerable mixing of the species throughout most of the rest of the transect. Cornel et al. (2003) studied some of these same localities, as well as similarly spaced localities in South Africa. They found that allozyme loci in the hybrid zone in California were in Hardy-Weinberg proportions, while loci from South Africa (where *pipiens* s.s. and *quinquefasciatus* overlap) show a Wahlund effect, indicating nonrandom mating.

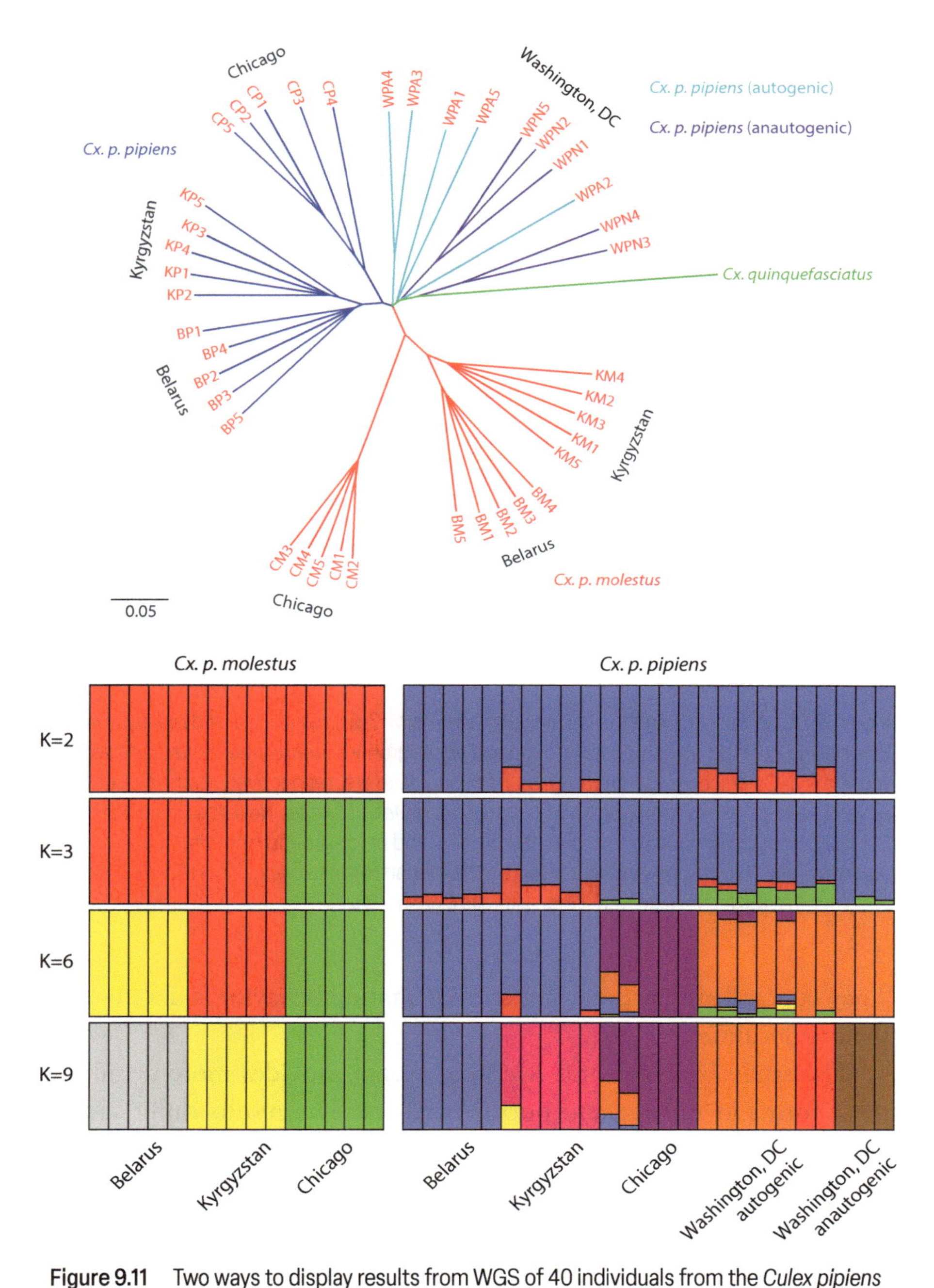

Figure 9.11 Two ways to display results from WGS of 40 individuals from the *Culex pipiens* complex. **Top**. A neighbor-joining tree. **Bottom**. An ADMIXTURE plot, with various values for K. From Yurchenko et al. (2020).

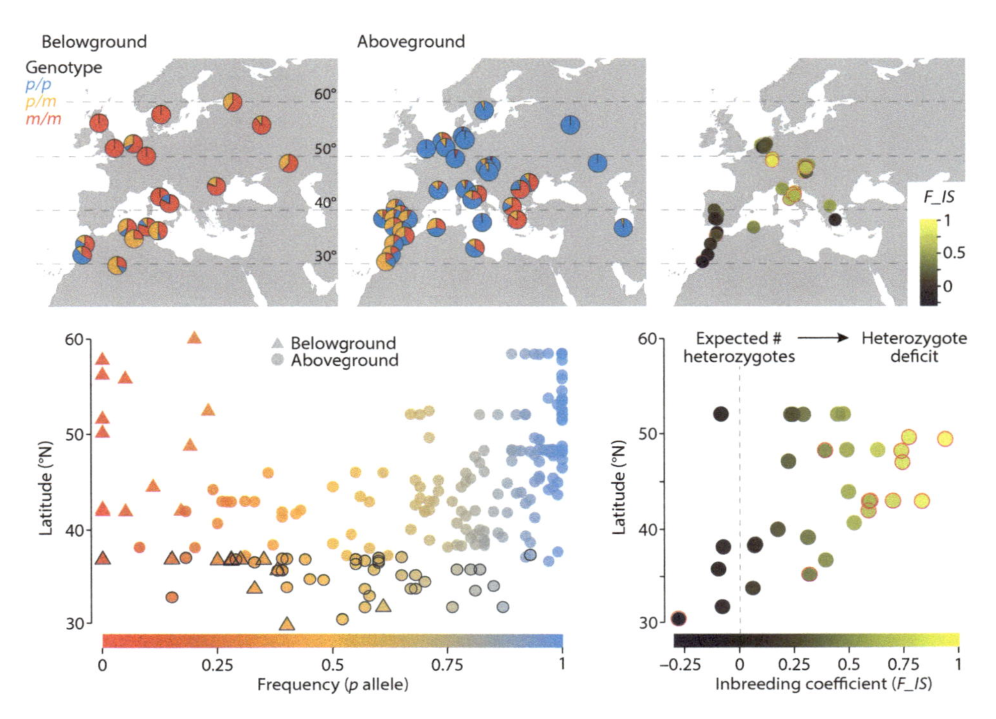

Figure 9.12 Maps showing the latitudinal gradients of *Culex pipiens* s.s. and *molestus* in Europe and northern Africa. Belowground and aboveground refer to where the collections were taken. Color-coded genotypes *p/p*, *p/m*, and *m/m* are, respectively, homozygous for *pipiens* alleles, heterozygous, and homozygous for *molestus* alleles. Note that in northern Europe, the aboveground (predominantly *pipiens*) and belowground (predominantly *molestus*) genotypes mix very little, while on the Iberian Peninsula and in northern Africa, considerable mixing is observed. From Haba and McBride (2022).

They concluded that these two species' mating behavior is different in the two localities.

The study by Haba and McBride (2022) provides more detail on these forms in Europe and northern Africa. The results differ somewhat from the two studies just discussed with regard to *pipiens* s.s. and *molestus*. The degree of mixing in the forms varies latitudinally. The northern collections are more distinct, with increased mixing when proceeding south (figure 9.12). But, consistent with the previous work, aboveground collections tend to have more mixing than those collected belowground.

While almost all population genetics studies use data from as many loci as possible, sometimes focusing on a few or even a single locus can lend insights. Chevillon et al. (1998) concentrated on a gene designated Aa-tA (a transaminase), which had previously been implicated in promoting autogeny by producing more efficient amino acid turnover, thus permitting a better uptake of nutrients at the larval stage. This allowed egg production without a blood meal. The frequency of this gene is significantly higher in belowground samples (presumably *molestus*, although this taxon was not yet recognized at the time of the study). Four other random allozyme loci, used as controls, were studied in the same samples and displayed no such pattern, adding credence to this gene having a causal effect on autogeny.

Notes

1. The quality of whole genome assemblies is assessed by the length of contiguous segments (contigs) of gene sequences, the total size of the assembly (does it match the measured size of the genome?), and what is call a BUSCO score—the proportion of conserved genes found in the assembly (Waterhouse et al. 2018).

2. STRUCTURE plots are a Bayesian statistical method to illustrate the genetic composition of population samples, first developed by Pritchard et al. (2000). A user specifies the number of genetic subdivisions assumed to be present (K), and the program separates the samples as efficiently as possible to make that many panmictic populations. Each individual in the sample is a line on the plot, with colors indicating the proportion of the loci assigned to each subgroup. A single color indicates all loci support assignment to one group, whereas individuals displaying mixed ancestry have bars of various color, indicating the relative contribution to the genome from each group.

3. Mitochondrial DNA (mtDNA) has also been studied for *aegypti* population genetics. Most of these data, however, should be viewed with caution, as it has been established that nuclear copies of the mitochondrial genome are ubiquitous in this species, confusing the issue of whether a cytoplasmic or nuclear copy has been assayed (Black and Bernhardt 2009; Hlang et al. 2009). The results are clear only when extreme care is taken—such as when multiple mtDNA genes are used and they co-vary; flanking sequences do not show nuclear DNA; or mtDNA is physically isolated by CsCl centrifugation. Paupy et al. (2012) provide an example of the accurate use of mtDNA.

4. In population genetics, time is measured in generations. Converting generations to absolute times requires assumptions about the number of generations per unit of time. Temperature and the availability of water are two factors controlling this number,

and they vary considerably in the distribution of *aegypti*. Generations per year likely vary from a few to 15.

Also, it is important to note that all such estimates of timing events based on genetic patterns make the assumption that the genetic variants being studied are selectively neutral, with random genetic drift being the major force changing allele frequencies over time. For the majority of DNA sequencing variants, this assumption is probably valid. Even in those subject to selection, this is likely to be weak. Thus if one analyzes thousands of SNPs, even if a few have undergone selection, they do not greatly bias the time estimates.

5. For various reasons, presence/absence information for a mosquito in a particular area can be inaccurate. It depends on how extensive and intensive the monitoring programs are, which vary considerably. Even if a strong monitoring program is in place, politics and economics may affect the reporting. Countries in which tourism is a big part of their economy are reluctant to be placed on lists of places with vectors of serious diseases.

6. The manner in which inversions in *gambiae* are designated is to use uppercase letters to indicate the chromosome arm, and lowercase letters to designate the inversion. The superscript $^{+}$ indicates an arbitrarily designated wild type or an uninverted sequence (+). Thus 2Rd indicates the presence of the d inversion in the right arm of chromosome 2, and $2Lb^{+}$ designates the uninverted gene order in the region covered by the b inversion in the left arm of chromosome 2.

7. It may be counterintuitive to conclude such patterns could arise by chance, but Endler (1977) showed that, theoretically, populations at the extreme ends of clines may have become fixed for different alleles (gene arrangements)—solely through random drift—when intermediate populations had not yet existed. Intermediate populations could slowly be founded by new immigrants from each end, resulting in gene frequency clines.

8. Selective sweeps can be detected in DNA sequencing data in places where one site has recently been under strong positive selection and risen rapidly in its frequency. Due to physical proximity on the chromosome, linked sites are dragged along. The result is that the region around the site undergoing a rapid increase in frequency has lower heterozygosity, due to greatly reduced frequencies (or even the elimination) of alleles not on the haplotype with the new, favorable mutation.

CHAPTER 10

Speciation

> Of course I allude to that *mystery of mysteries*, the replacement of extinct species by others. Many will doubtless think your speculations too bold—but it is as well to face the difficulty at once.[1]
>
> —John Herschel, 1836

Since the inception of evolutionary biology, the issue of speciation (how one evolutionary lineage bifurcates into two) has been a deceptively complex issue to fully comprehend. Even after 150+ years of research we have only a partial understanding. A situation much prized by those studying the speciation process is to identify species that are at or very near the point of splitting, what Dobzhansky (1935) called *status nascendi*, or the birth of a new species. At least two of our three species complexes offer an opportunity for such studies.

Researching the origin of new species requires making explicit definitions of what is to be studied. The first is to define species. Here I will adopt the biological species concept mentioned in chapter 1. The advantage of this concept is that it has a well-stated, empirically approachable definition: the cessation of gene exchange. This may take the form of non-mating, due to behavioral, ecological, or geographic isolation. If mating does takes place, the hybrid embryos do not develop into adults, or, if they do, they have low fitness (e.g., sterility). From a broader perspective, species recognized under the BSC are *in-*

dependent evolutionary units—that is, genetic changes in one do not directly affect genetics of other species. Exceptions occur when introgressive hybridization occurs, which traditionally was thought to be rare. Recent evidence, however, especially from molecular studies, has revealed more examples of introgression than were previously assumed (Mallet et al. 2016; Edelman and Mallet 2021).

Multiple aspects of the speciation process can be addressed.

1. *Spatial considerations*: Did the genetic differentiation occur when the populations were sympatric, or is allopatry (geographically separate) or parapatry (geographically adjacent) always required?
2. *Nature of the genetic differences*: Are the nascent species becoming genetically isolated due to chromosomal changes, single genes, or multiple genes clustered or spread across the genomes? Or is it possible to have reproductive isolation in the absence of genome divergence, a possible example of which exists with these mosquitoes?
3. *Force(s) causing the genetic differentiation*: Is this due primarily to random genetic drift, selection, or new mutation(s)?
4. *Nature of the barrier(s) to gene exchange*: These are noted on the preceding page.

Aedes aegypti

Despite the considerable morphological, behavioral, and ecological diversity of this taxon, using the BSC, there is almost no evidence that *aegypti* is more than a single, widespread, polytypic species. There are taxa, however, that are very closely related, and at least one that is only partially reproductively isolated from *aegypti*. These occur on islands off the east coast of Africa.

Ae. mascarensis was originally thought to occur only on the island of Mauritius,[2] although it was recently collected on Réunion Island (D. Ayala et al., unpublished data). This species is morphologically distinct from *aegypti* in having more white scaling on the dorsal thorax, although this is quite variable (Hartberg and McClelland 1973). Crosses with *aegypti* produce viable, fertile F_1 males and females. In further generations (F_2, etc.) and backcrosses, however, male offspring display a large proportion of abnormalities, such as unrotated genetalia and intersex (male and female) characteristics (Hartberg and Craig 1970). This has been termed "hybrid breakdown." This phenomenon is consistent with multigenic, epistatic (interacting) sets of genes distinguishing the species that, when recombined in the F_1, produce gametes that give rise to abnormalities.

Ae. pia was described as a morphologically distinct species, but quite similar to *aegypti*, that was found on Mayotte in the southwestern Indian Ocean (Le Goff et al. 2013). *Ae. pia's* reproductive compatibility with *aegypti* and *mascarensis* remains to be determined. There is likely at least one more undescribed, closely related species on these islands, as evidenced by a lineage on Madagascar that is morphologically *aegypti* but genetically more distant from *aegypti* than is *mascarensis* (Soghegian et al. 2020).

The important point is that there are poorly studied taxa that are closely related to *aegypti* in the ancestral range of southwestern Indian Ocean islands that may be partially reproductively isolated and represent early stages in the speciation process. Given the advantages of *aegypti* for lab-based research, this is potentially a very favorable opportunity to study speciation in mosquitoes.

Anopheles gambiae

Relationships

It is only recently that some degree of understanding of the complicated relationship of members of this complex has been elucidated, primarily through extensive whole genome sequencing. The underlying confusion is due to the multiplicity of genetic and ecological factors involved. Inversion polymorphisms are ubiquitous, and WGS has clearly shown that inversions, although monophyletic, are retained during speciation as well as passed across species boundaries—that is, they display introgression. So, while inversions were initially used to define subdivisions (e.g., "chromosomal forms"), this is misleading. Introgression is also heterogeneous across the genomes and may vary geographically.

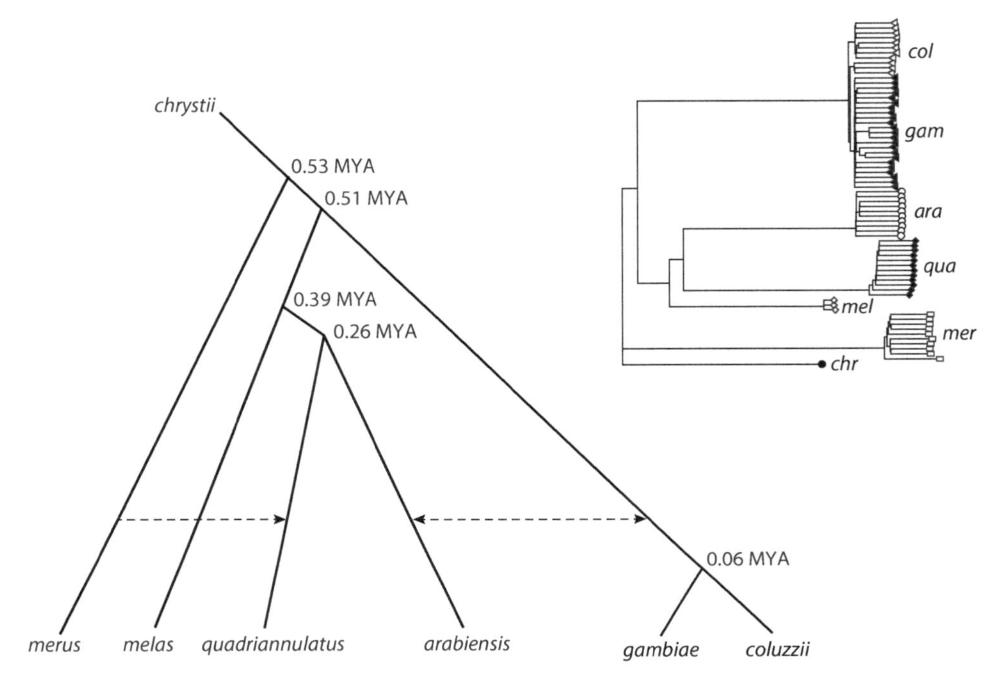

Figure 10.1 **Left**: Phylogenetic trees of the *Anopheles gambiae* complex, based on whole genome sequencing. Independent analyses by Fontaine et al. (2015) and Thawornwattana et al. (2018) produced identical branching patterns. Estimated dates for the nodes are from the latter paper. The dotted lines indicating introgression are for inversions (3La on the left and 2La on the right). **Right**: Multiple strains of each taxon form cohesive monophyletic groups. From Fontaine et al. (2015).

Figure 10.1 shows the consensus species relationships among six of the named species of the *gambiae* complex, based on WGS (Fontaine et al. 2015; Thawornwattana et al. 2018). The upper right tree indicates that when multiple individuals of each named taxon are sequenced, they form monophyletic cohesive groups, confirming the genetic distinctiveness of each species. It is only since robust trees like this have been available that we can begin to understand the speciation process.

While this is a consensus tree, not all parts of the genome support this topology. Gene trees based on subsets of the genome, especially the autosomes, are often at odds with this species tree. This is likely due to introgression. The X chromosome provides the clearest species

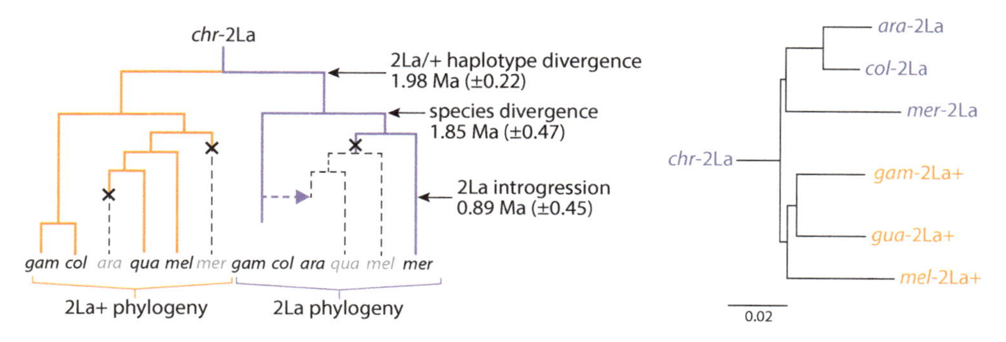

Figure 10.2 Phylogenetic trees showing different branching patterns, depending on sequences within breakpoints of the 2La inversion in the *Anopheles gambiae* complex. **Left**. The tree indicates that all species are fixed for either 2La or 2La⁺, except for *coluzzii* and *gambiae*, which are polymorphic for these arrangements. **Right**. The tree indicates that these two arrangements are each monophyletic. Thus sharing by *coluzzii* and *gambiae* s.s. most likely is due to introgression. Both trees are from Fontaine et al. (2015).

tree, with less evidence of introgression (Crawford et al. 2015; Fontaine et al. 2015).

Gene trees based on DNA sequences between the breakpoints of the inversions are especially insightful. Figure 10.2 shows the inversion trees for the 2La inversion. The 2La arrangement is ancestral to 2La⁺ being present in the outgroup *chrystii* (chr). While the two alternative arrangements are each monophyletic (figure 10.2, right), sequences within the alternatives support distinctly different trees (figure 10.2, left). The dissimilarities between the trees can be resolved by postulating that 2La was passed (introgressed) between *arabiensis* and the clade leading to *gambiae*/*coluzzii*. Similar analyses led to evidence of 3La introgressing from *merus* to *quadriannulatus* (figure 10.1, left).

These processes, inferred from DNA sequencing including WGS analyses (e.g., Besansky et al. 2003), are consistent with lab studies on the fate of subsections of the genome in hybrid populations between *gambiae* and *arabiensis*. Slotman et al. (2005) found that microsatellites on X and 3R decreased rapidly in hybrid populations, whereas alleles on 2L persisted at their original frequencies for up to 19 generations. This indicates selection against introgression for much of the genome, but not for 2L. This is also consistent with the X chromosome being less vulnerable to introgression and thus is informative in constructing species trees. Introgression of only parts of the genome (selective introgression) is tolerated.

Figure 10.1 (left) also shows time estimates for the splitting events. Given the lack of fossils, some form of a molecular clock is invoked in designating times. These involve various assumptions (chapter 9, note 4), so the times shown are truly estimates. The most recent split is between *coluzzii* and *gambiae* s.s. Thus these two provide the best evidence for being taxa in *status nascendi*.

The Speciation Process

While the above details of *gambiae* s.l. relationships and introgression have only been elucidated recently, Coluzzi (1982) had long inferred an important role for inversions in the speciation process. The core postulate in what he called "chromosomal speciation" is that inversions, by tying together sets of genes that confer special adaptive properties, allow species to expand their distributions into marginal habitats. A more general model of speciation emphasizes the role of epistatic gene combinations, held together by reduced recombination. Paracentric inversions are one effective means of doing so, and tight linkages or concentrations in regions of low recombination may also play a role. This model has been most clearly explicated by Small et al. (2023) for another African malaria vector, *funestus*. Given our present understanding, the *gambiae* complex fits this model well.

In the case of *coluzzii* and *gambiae* s.s., their main ecological difference occurs in larval breeding. *An. gambiae* s.s. predominates in temporary rain-filled pools, whereas *coluzzii* larvae are most often found in more permanent man-made water, such as rice fields (Lehmann and Diabate 2008). Rice cultivation in sub-Saharan Africa began about 2,000 years ago. Larval predator avoidance represents a major adaptation in these two taxa. In permanent pools that have more predators than temporary pools, *coluzzii* outcompetes *gambiae* s.s., whereas in predator-free pools, *gambiae* s.s. outcompetes *coluzzii* (Diabate et al. 2005, 2008).

In comparing whole genomes of *coluzzii* and *gambiae*, it is clear that the greatest differences are between the breakpoints of the 2La inversion (Lawniczak et al. 2010; Crawford et al. 2015; Fontaine et al. 2015), implying that genes in this region were important in their expansion into the new rice field niche, although perhaps this is not the only genomic region involved. An important point in the recentness of this expansion is that there has not been time for new gene variants to arise from mutations. Rather, some new combination of preexisting variants arose to allow this expansion. One observation, first documented in *Drosophila*, is that inversion polymorphism decreases toward the margins of species distributions. Carson (1959) postulated that this allows more adaptive flexibility at the margins, due to a higher frequency of recombining inversion homozygotes, thus shuffling alleles at epistatically interacting genes to produce novel adaptive properties.

Once a species is established in a new niche, there will be selection to retain the critical block of genes and keep it from being swamped or destroyed by recombination from the larger ancestral group. Thus variants promoting reproductive isolation between the new niche and their ancestral habitat will be favored. These may reside either in the initial block of genes (e.g., inversion) or elsewhere in the genome, where recombination is reduced (e.g., near centromeres). This appears to be the case for the *gambiae* complex, where highly diverged "speciation islands" near centromeres were documented, independent of inversions (Turner et al. 2005; B. White et al. 2010).

In the *gambiae* complex, the introduction of the molecularly diagnostic M and S forms provided a convenient tool for studying reproductive isolation between sympatric populations. In this complex, premating isolation, due to the spatial separation of mating swarms, is a common mechanism to reduce gene exchange (Lehmann and Diabate 2008). Specifically, with respect to *coluzzii* and *gambiae*, two differentiated islands near the X and 2L centromeres were used to study mating swarms in western Burkina Faso (Niang et al. 2022). Carriers of alternative X chromosome islands perfectly separated spatially into swarms corresponding to *coluzzii* and *gambiae*. Laboratory crosses reveal no post-mating isolation between these taxa (Tripet et al. 2005; Diabate et al. 2007).

The importance of temporary versus more permanent larval breeding sites also applies to cases other than those in rice fields (Lehmann and Diabate 2008). A parallel situation has been studied in Mali with regard to laterite rock pools (enriched by iron and aluminum) subject to frequent swamping by rising water (Manoukis et al. 2009). A population, once designated "Bamako chromosomal form,"[3] is fixed for the 2Rj inversion, which is rare or absent in most *gambiae* s.s. populations. At a site near Banambani, 2Rj and $2Rj^+$ homozygotes represented 98.5% of the individuals collected. In other words, almost no heterozygotes were found, likely reflecting mating isolation between carriers of the two homozygotes. The frequency of 2Rj was higher in larvae from laterite rock pools, compared with more permanent puddles and swamps. Importantly, only molecular S forms were included in the study, providing strong evidence that this molecular marker does not represent a single species. Instead, there may be cryptic species within what is today recognized as *gambiae* s.s.

Pombi et al. (2017) reviewed much of the literature on the *gambiae* complex relating to reproductive isolation and reached intriguing conclusions. Pre-mating isolation exists in most instances of very closely related taxa in this group. Their review, however, also implicate post-mating isolation, in which hybrids between ecotypes have a fitness cost. A hybrid fitness deficit was inferred by the decrease in hybrids from the larval to adult stages. The more a species is ecologically divergent, the greater the post-mating fitness cost. The authors argued that this is the expected pattern in ecological speciation theory, as greater ecological divergence implies that hybrids would be at a greater disadvantage.

Culex pipiens

The *pipiens* complex comprises a number of taxa of sometimes ambiguous taxonomic status (chapter 1). The two most widespread are temperate *pipiens* s.s. and tropical/semitropical *quinquefasciatus* (see figure 1.10). Where they meet, hybrid zones are formed, which seem to be stable. References to much of the following can be found in the detailed review by Haba and McBride (2022).

More pertinent, however, from a speciation standpoint is the ecotype (or form) *molestus* that overlaps in distribution with *pipiens* s.s. Five traits were originally identified that distinguished this taxon from *pipiens* s.s. and *quinquefasciatus*: (1) larval breeding underground (in subways, basements, etc.), (2) mating in enclosed spaces (stegonomy), (3) mammal biting, (4) a lack of diapause, and (5) autogeny (Haba and McBride 2022). Form *molestus* can exist in sympatry with *pipiens* s.s. and *quinquefasciatus* yet remain distinct, especially in northern regions. In Europe, as one goes south, the distinction breaks down and intermediate hybrid populations exist (figures 9.12 and 10.3). Farther south, in northern Africa, hybrids are dominant.

How do these two distinct ecotypes remain genetically differentiated when they are sympatric, at least over much of their range? Mating behavior is the likely barrier to gene exchange. This is detected by a distinct deficit of expected heterozygotes when both forms are collected in the same trap. The study by Urbanelli et al. (1981) is particularly relevant in this regard. They reared large numbers of genetically distinct aboveground *molestus* collected in Rome and released them in a rural region where only *pipiens* s.s. were present. They then made collections of mating swarms at different heights: near the ground, and two meters up in foliage. Almost all of the ground-collected mosquitoes were *molestus*, and those higher up were *pipiens*. This situation remained stable over four months as collections continued.

Similarly, an outdoor cage experiment involving form *molestus* and *quinquefasciatus* (*fatigans*) demonstrated strong pre-mating isolation. Egg rafts were collected over a 13-week period when both types were in the cage. Only 1 egg raft out of the 66 examined was hybrid. This implies that form *molestus* is more behaviorally isolated from *quinquefasciatus* than from *pipiens* s.s., given the frequencies of hybrids where *pipiens* and form *molestus* are sympatric (e.g., figure 10.3).

With regard to speciation, two other findings in this complex are potentially of interest. The hybrid origin of *pallens* in northeastern Asia is unusual, and perhaps unique for mosquitoes. The likely populations of *pipiens* s.s. and *quinquefasciatus* giving rise to *pallens* have recently been identified (Haba et al., submitted).

Another unusual observation in the complex is the potential role of

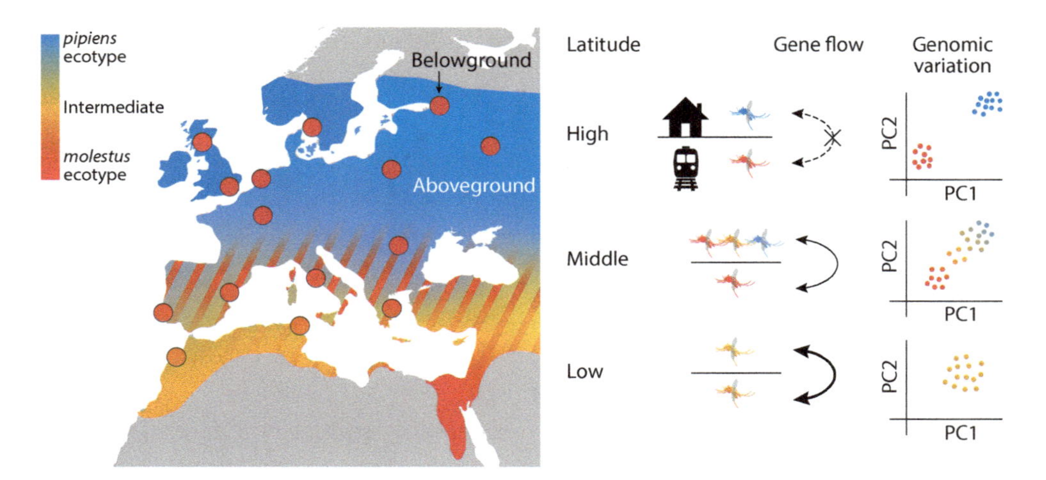

Figure 10.3 Summary of latitudinal gradients of reproductive isolation in the *Culex pipiens* complex. From Haba and McBride (2022).

endosymbionts in reproductive isolation. In 1967, Hans Lavan reported that crosses between geographic populations of some *pipiens* were sterile—that is, no viable offspring were produced. Yen and Barr (1971, 1973) subsequently showed that this incompatibility is maternally inherited through egg cytoplasm, with the causative agent being a symbiotic bacteria, *Wolbachia pipientis*. The infertility between infected and uninfected strains can be cured by treatment with antibiotics, thus confirming that reproductive incompatibility is independent of the host genomes. Since these early observations, a considerable body of literature has arisen examining this system, including theoretical studies (Turelli et al. 2022 and references therein) and molecular biology (Hochstrasser 2023 and references therein).

There is no evidence yet that *Wolbachia* infections are playing a role in speciation in this complex, although further theoretical and empirical studies may implicate this endosymbiont in speciation. This would imply a novel possibility—that reproductive isolation may arise without genetic differentiation and be caused just by infection with a microbial symbiont. (The use of *Wolbachia* in control programs is discussed in chapter 11.)

Notes

1. The phrase "mystery of mysteries" is often ascribed to Darwin, but the precise quote in *The Origin of Species* where he uses this phrase is an attribution: "These facts seem to me to throw some light on the origin of species—that mystery of mysteries, as it has been called by one of our greatest philosophers." The philosopher Darwin praised so highly was John Herschel, a true polymath who made seminal contributions to astronomy, mathematics, botany, and photography, as well as being the inventor of the blueprint. See Cannon (1961) for more details.

2. The description of how MacGregor found *mascarensis* is quite delightful: "Soon after my arrival in Mauritius, while I was in the Botanical Gardens at Pamplemousses one morning, I noticed a small black and white mosquito buzzing around my legs. Its flight was rapid, and it seemed easily frightened; consequently it disappeared as soon as I attempted to capture it. I noticed, however, that the insect seemed to have a conspicuously white thorax" (MacGregor 1924). Try using this kind of writing with today's fusty scientific journal editors!

3. Near the end of Mario Coluzzi's distinguished career, it became clear that if any new species in the *gambiae* complex was formally described, it should be named in his honor. In 2011, Charles Taylor and I were visiting him in his home in Rome, where he was confined in his waning years, and asked which of the then-recognized chromosomal forms of the *gambiae* complex he would like to see named after him. Coluzzi's response was Bamako, as he felt this had the best chance of standing up to further scrutiny and the least chance of being abandoned as more information was accumulated. Mopti (M molecular form) was chosen (Coetzee et al. 2013).

CHAPTER 11

Microbiomes, Diseases, and Innate Immunity

Biology's sleeping giant

—Carl Woese, 1998

Microbes play a large role in multiple facets of mosquito biology, and we are just beginning to understand its extent. This chapter will present aspects of mosquito-microbe interactions, including the microbes transmitted by our three mosquitoes that cause human diseases. This is followed by a very brief discussion of how mosquitoes fight off microbial infections.

Microbiomes

The unseen microbial world is largely *terra incognita* or a "sleeping giant," as dubbed by one of the pioneers of this unknown territory, Carl Woese (1998). Until about 25 years ago, what we knew about microbes came from culturing them in the lab, but this represents only a very small fraction of their total diversity; most microbes cannot be cultured and thus have remained totally unknown. Today, DNA extracts from various environments (soil, water, skin, guts, etc.) produce hundreds or thousands of unique sequences from microbes never seen before. These assemblies are called "microbiomes." Microbiomes are studied by using sets of PCR primers that amplify bacterial, fungal, and/or algal genomes. The output is then compared with a large database of known microbes. Results are usually a very long list of taxa, for which little is known from a functional standpoint. Dada et al. (2021) have made recommendations for standardizing microbiome research on mosquitoes.

Microbes are important sources of food for larvae (chapter 4), so it is not surprising the larval gut microbiome reflects the microbiome of the larval water. In a field study, Coon et al. (2016) showed that the microbiota of three species of mosquitoes (*aegypti*, *quinquefasciatus*, and *albopictus*) collected in the same larval water had similar microbiomes, but the microbiomes varied across different sites. Although microbes are not found inside eggs (except for endosymbionts), microbiomes can be transferred across generations on the surface of eggs, the shells of which may be ingested by larvae (Vinayagam et al. 2023). Thus, to some degree, microbiomes are "inherited." Upon pupation, the larval gut microbiome is ejected, and pupae are almost sterile, so the resulting adults initially have few microbes (Moll et al. 2001). Adults pick up microbes from ingesting water, as well as from nectar (Zouache et al. 2022). Different organs in mosquitoes have been found to harbor different sets of bacteria (Mancini et al. 2018).

Microbiomes have multiple effects on mosquitoes, including affecting vector competence (reviews in Cansado-Utilla et al. 2021; Vinayagam et al. 2023). The information available to date is largely correlative, rather than providing any mechanistic understanding of how nonpathogenic microbes affect a female's ability to transmit pathogens. Considerable work, however, has been done in attempting to use mosquito-associated microbes to control disease transmission. Two microbes have been studied in some detail and deserve discussion here.

Wolbachia pipientis is an endosymbiont first identified in *pipiens* that can cause sterility in crosses (chapter 10). *Wolbachia* can also be found in *Drosophila*, where it can shorten their lifespan. McMeniman et al. (2009) artificially infected *aegypti*—which does not naturally have *Wolbachia* (Gloria-Soria et al. 2018a)—and found a similar decrease in adult survival. More surprisingly, infected *aegypti* females had a much lower capacity to transmit pathogenic viruses and malaria (Moreira et al. 2009). *Wolbachia* evidently stimulates the female's innate immune system, suppressing the replication of the pathogens (Rances et al. 2012; Pan et al. 2018). This has led to a widespread effort to introduce *Wolbachia* into natural populations of *aegypti* to control diseases (Ross et al. 2019). Females carrying *Wolbachia* have an advantage over noninfected females, as the former are fertile with both infected and noninfected males. Because of this, *Wolbachia* can self-spread in a population. Successful reductions of dengue cases have been observed in a number of locations where *Wolbachia*-infected *aegypti* have reached at least 80% of the local *aegypti* population (references in Powell 2022). The sterility effects of *Wolbachia* in *aegypti* have also been used to suppress population numbers (Crawford et al. 2020; Beebe et al. 2021).

Paratransgenesis (modifications using microbes naturally found in mosquitoes) can be used to kill or inhibit pathogens. A particularly promising bacterium in this regard is *Asaia*, an abundant bacteria found naturally in several mosquitoes, including *gambiae* and *aegypti* (Chouaia et al. 2010). It is easily cultured and can be genetically manipulated using standard technologies (Favia et al. 2007; Damiani et al. 2010; Mancini and Favia 2022). It is found in a mosquito's midgut and salivary glands—key tissues for pathogen transmission. Its abundance in females is not affected by presence of *Plasmodium* and generally seems to be resistant to the female's innate immune system (Capone et al. 2013). While paratransgenesis has potential as a means of reducing disease transmission, this has yet to be fully demonstrated for *Asaia* or any other microbe.

Diseases

The major diseases transmitted by each of our three mosquitoes fall into three distinct classes: viruses (*aegypti*), protozoan *Plasmodium* (*gambiae*), and filarial nematode worms (*pipiens*). While these are the major pathogens transmitted to humans by each taxon, all three mosquitoes can transmit all three types of pathogens, although less frequently and seldom involving transmission to humans. Common features of pathogen transmission include its acquisition through a first blood meal; migration of the pathogen from the gut to the salivary glands (for viruses and malaria) or proboscis (for nematodes); and transmission to a new host by a second blood meal (figure 11.1). Details differ, however. Powell (2019) and Fontenille and Powell (2020) consider the evolution of these three-species systems (mosquito, pathogen, and vertebrate).

Viruses

Viruses that are arthropod borne are called "arboviruses." Three major viral diseases are transmitted by *aegypti*: yellow fever, dengue, and chikungunya.[1] All are RNA viruses. The yellow fever virus (YFV) and dengue virus (DENV) are in the family Flaviviridae, while the chikungunya virus (CHIKV) is an alphavirus in the family Togoviridae. While other mosquitoes can transmit these viruses, *aegypti* has the highest vector competence, likely due to a common origin of the virus and this mosquito in Africa (Powell 2018b). Variation in the genomes of these viruses is considerable, which is not surprising, as RNA genomes have a high rate of mutation. Dengue has at least four distinct serotypes (microorganisms with a common set of antigens) that produce different immune reactions in humans—that is, immunity to one serotype does not confer immunity to the others. In fact, after becoming immune to one serotype, when infected with a different serotype, the disease is often much worse and more often leads to fatal hemorrhagic fever.

Figure 11.1A shows the cycle of a virus through a mosquito. The time it takes to migrate from a blood meal to the salivary glands is the extrinsic incubation period, or EIP. This varies considerably, depending on virus strain, temperature, and the strain of the mosquito. For example, for YFV, EIP varies from 4 days at 37°C to 18 days at 21°C, and for DENV, from 10 days at 30°C to 35°C to 25 days at 25°C (Davis 1932; Chan and Johansson 2012). For CHIKV, it is less clear if EIP is temperature dependent (Mbaika et al. 2016). The relative effectiveness of *aegypti* in transmitting these viruses varies greatly among strains from different localities (Souza-Neto et al. 2019).

While this is the usual manner of transmission, these viruses can also be transmitted through an egg's cytoplasm (transovarial transmission, also known as vertical transmission), thus sustaining the infection across

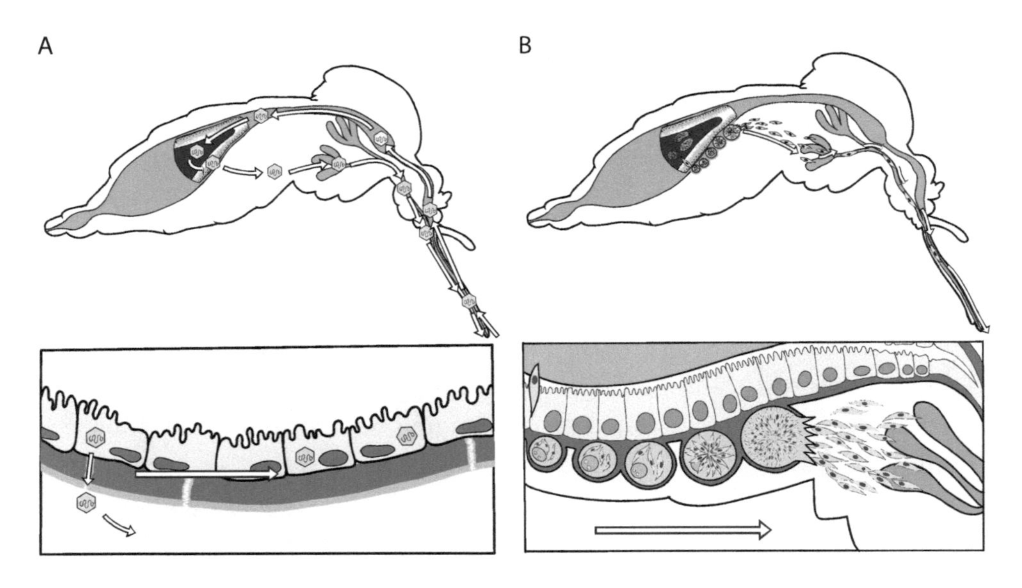
A
B

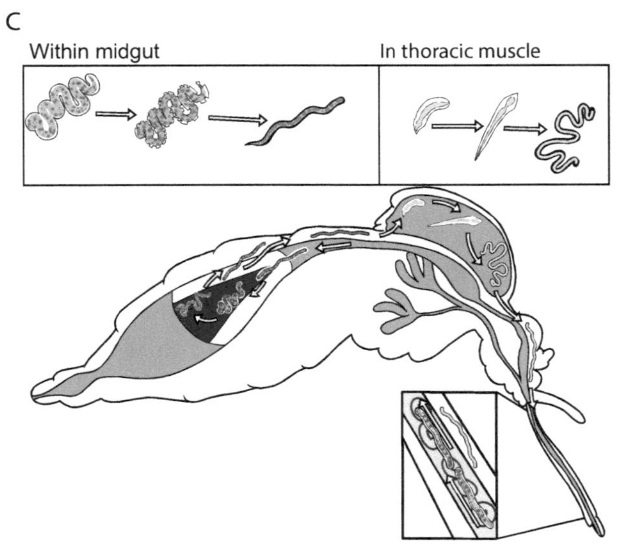
C
Within midgut
In thoracic muscle

mosquito generations (Beaty et al. 1980; Joshi et al. 2002; Heath et al. 2020). While most demonstrations of transovarial transmission have taken place in the lab, this has also been observed in nature (Khin and Than 1983; Lee and Rohani 2005; Heath et al. 2020).

Another flavivirus, West Nile virus (WNV), was first identified in 1937 in Uganda (Smithburn et al. 1940). This arbovirus is distinct from those described above in being primarily transmitted among birds via various *Culex* species (Rochlin et al. 2019). *C. pipiens*, which takes blood meals from both birds and humans (chapter 6), can act as a bridge vector, bringing this avian pathogen to humans. Humans are a "dead end" host, as levels of this virus are not high enough in humans to be transmitted by mosquitoes. Given the widespread distribution of *Culex* (*pipiens* s.l. in particular), WNV has a broad distribution, including tropical and temperate regions. WNV was first reported in the United States in 1999, likely imported by birds in a New York City zoo (Lanciotti et al. 1999). The original introduced genotype of WNV in the United States has largely been replaced by an evolved virus that has increased transmissibility in American crows (Bialosuknia et al. 2022).

Figure 11.1 (*opposite*) Passage of pathogens through mosquitoes. **A**. Viruses. Viruses are ingested during a female mosquito's blood meal (**top**) and are deposited into the midgut, which expands during ingestion of a blood meal. The virus then infects the midgut cells and undergoes replication. The virus escapes the midgut (**bottom**) into the hemolymph through the basal lamina, then travels to the salivary glands, and is excreted during subsequent blood meals. **B**. Malaria. Female *Anopheles* mosquitoes ingest gametocytes (**top**) during blood feeding. Male and female gametocytes mate in the mosquito's midgut, forming a fertilized egg (zygote). The zygote enlarges and becomes motile, forming an ookinete, which burrows through the midgut walls to form an oocyst on the exterior surface (**bottom**). Thousands of sporozoites develop over a period of 8–15 days within the oocyst. The oocyst will then burst and release the sporozoites into the mosquito's hemolymph, where they travel to the salivary glands. Sporozoites are transmitted to the next host via a subsequent blood feed. **C**. Filariasis. Female mosquitos ingest sheathed microfilariae (**bottom**) with an infected blood meal. Microfilariae unsheathe within the midgut (**top left**) and then exit the midgut and migrate to the thoracic muscles. Within the thoracic muscles, the microfilariae develop into first-, second-, and then third-instar infectious larvae (**top right**). The third-instar infectious larvae exit the thoracic muscles and migrate to the proboscis sheath, where they are then deposited into the next blood meal host. Illustrations by Jacquelyn LaReau.

Illness due to a WNV infection in humans is relatively mild, with about 80% of affected humans being asymptomatic.

Malaria

Plasmodium has a much more complex life cycle than viruses (figure 11.1B). It has both a sexual and asexual stage. Haploid gametocytes from the vertebrate host's blood are taken up in a blood meal and develop into male and female gametes in the midgut of a mosquito, where fertilization occurs, producing diploid ookinetes that traverse the midgut wall. This all takes about one day. While attached to the outside of the midgut wall, ookinetes develop into mature oocysts that produce thousands of haploid sporozoites by meiosis. These are released into the hemolymph and infect the salivary glands. This takes on the order of 12–14 days. At most stages, the population of *Plasmodium* cells numbers in the hundreds to thousands, except at the stage when traversing the midgut. Very few ookinetes (generally about five) make it through (Sinden 2017).

There are five species of *Plasmodium* that infect humans: *falciparum*, *vivax*, *malariae*, *ovale*, and *knowlesi*. *P. falciparum* dominates in sub-Saharan Africa and is the most fatal of the five. The major bird malaria parasite that has taken such a toll on Hawaiian bird fauna is *P. relictum*.

Filariasis

Nematode worms causing filariasis differ from arboviruses and malaria in a number of ways. Microfilariae ingested in a blood meal shed their sheaths, penetrate the gut wall, and migrate to the thoracic muscles, where they grow to a third instar stage (figure 11.1C). These then migrate directly to the proboscis. When another blood meal is taken, the worms are deposited on the host's skin or just under the skin, rather than into capillaries.

Human filariasis is mainly caused by two species of nematodes, *Wuchereria bancrofti* and *Brugia malayi*. In humans, the worms lodge in the lymphatic system, causing lymphatic filariasis, sometimes leading to extreme swelling (elephantiasis).

While members of the *pipiens* complex, especially *quinquefasciatus*, are the most widespread and major vectors of filariasis, filarial worms can be transmitted by a larger range of mosquitoes than most mosquito-borne pathogens. In the South Pacific, several *Aedes* are vectors, and in Africa, *An. gambiae* s.l. and *funestus* (chapter 12) are major vectors. Bockarie et al. (2009) present more details.

Infection and Mosquito Fitness

Do mosquitoes suffer a fitness deficit when infected with these pathogens? In the case of the viruses mentioned above, there is no evidence that *aegypti* females carrying them have decreased fitness, although this has not been widely studied. The fitness cost to anophelines that host *Plasmodium* infections has been more thoroughly investigated. When combinations of mosquito species and *Plasmodium* species that occur naturally are tested in the lab, there is little or no evidence of reduced fitness (Ferguson and Read 2002). There is some evidence, however, of competition for nutrients between this parasite and mosquito tissues (Hajkazemian et al. 2021). Very few such studies have included *gambiae*, largely due to its difficulty of being maintained in the lab. Even fewer studies have examined the effect of filarial worms on *pipiens*, although what has been done indicates little effect (Hu 1939; Lima et al. 2003).

Note: This is just a very brief introduction to the *major* diseases spread by mosquitoes—that is, those causing the most illnesses and deaths both today and throughout history. This book, focusing on the biology of three mosquitoes, does not deal with the myriad of other pathogens transmitted by a number of other mosquitoes, nor with medical issues or efforts to control these diseases.

Innate Insect Immunity

Until about 40 years ago, insects were assumed to not have an immune system. Beginning about 1980, considerable research has documented a rather sophisticated innate immune system in insects, including mosquitoes. At least some of this interest is due to the possible role immunity may play in mosquitoes' abilities to transmit human pathogenic microbes. The immune system in insects is innate, in contrast to acquired immunity. In other words, insects have a fixed set of genes and

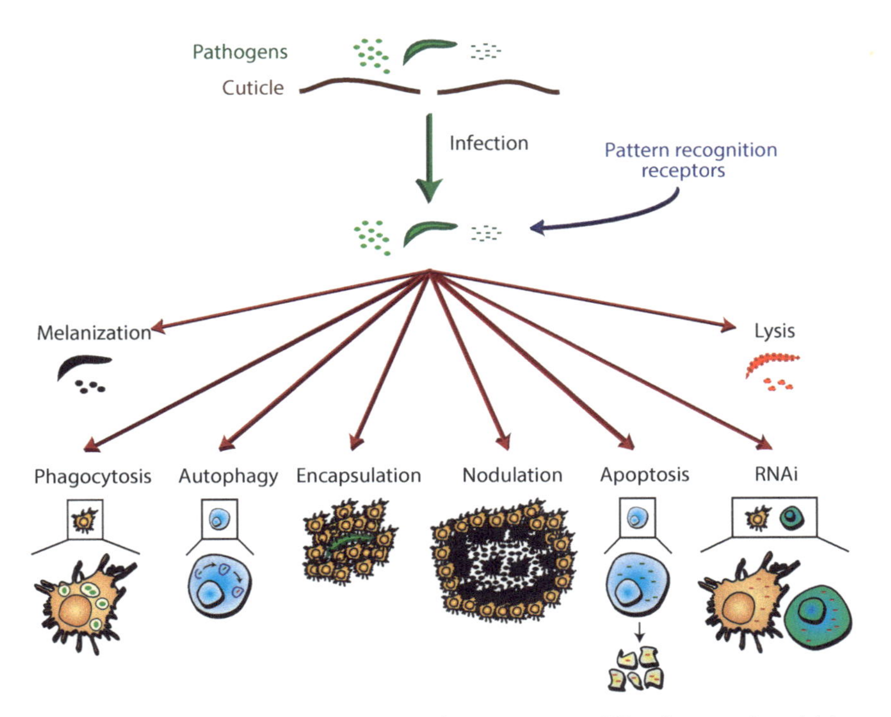

Figure 11.2 Immune effector mechanisms of insects. Insects kill pathogens via multiple immune strategies. From Hillyer (2016).

pathways that can act to fight off infections, although the expression levels of these genes can be greatly altered by environmental factors, including infections with microbes.

There are several known pathways that insects use to fight off microbial infections (figure 11.2). Each has pattern-recognition receptors on the membrane surface of several tissues: fat bodies, the gut lining, hemocytes, and the like. Once activated, the pathway signals the nucleus to produce antimicrobial peptides that can kill the invading microbe. In addition, hemocytes can activate a melanization process that encapsulates the invader with black melanin. The details of these processes are quite involved. General reviews include Hillyer (2016) and Bhattacharjee et al. (2023). For arboviruses, see Prince et al. (2023), and for malaria, Pohl and Cockburn (2022).

These innate immune systems are favorite targets for attempts to use genetic manipulations to produce mosquitoes refractory to the transmission of human pathogens. Dong et al. (2022) and Kefi et al. (2024) review such attempts for malaria.

Note

1. It is questionable whether Zika (ZIKV) is a distinct arbovirus, as it is very closely related to DENV and can be considered a variant of dengue (Gaunt et al. 2001). It was this variant that swept through Brazil and other parts of South America in about 12 months in 2015–16, causing considerable alarm and publicity. At the time, the methods used to distinguish the Zika variant from other dengue types were unreliable. Since 2017, as better detection methods were put in place, ZIKV has almost disappeared from much of the territory it supposedly invaded. Except for reports of increased microcephaly in babies of mothers with Zika, infections with this variant were on the mild side for dengue. And there are questions as to whether there was any cause and effect between the virus and microcephaly, as the data are murky. As one example, midway through the outbreak, Brazilian authorities changed the definition of microcephaly to include more babies, irrespective of any association with Zika infections.

CHAPTER 12

Other Mosquitoes

While the focus of this book has been on the three best known mosquitoes, it would be remiss not to at least introduce other species that deserve attention. It is reasonable to ask whether, if our three mosquitoes were all eliminated or rendered incapable of transmitting human pathogens, would mankind finally be free of mosquito-borne diseases? Not by a long shot! Plus, mosquitoes have been research subjects for reasons other than their role in disease transmission, including insect physiology, neurobiology, ecology, and behavior.

Impacts on Human Health

Aedes

The obvious candidate for the second-most-important member of this genus is *albopictus*. For humans living at higher latitudes, *albopictus* is of more concern, as it has the ability to overwinter in diapause. Figure 12.1 shows its distribution as of early 2022, with the expectation that its spread is not yet over. Of particular note, temperate regions in North America, Europe, and Asia that are free of *aegypti* have this mosquito

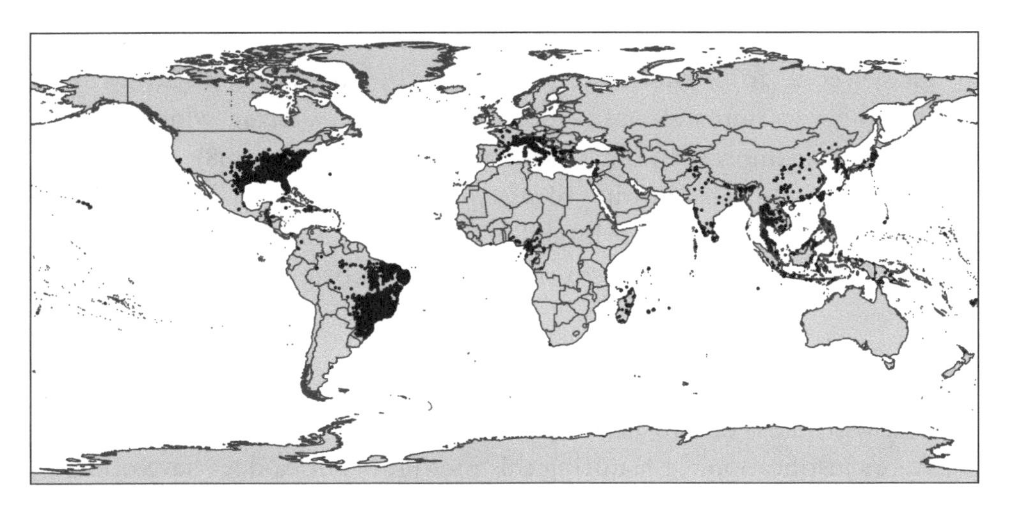

Figure 12.1 Distribution of *Aedes albopictus* (indicated in black), as of January 2022. From Laporta et al. (2023).

vector (compare figure 12.1 with figure 1.6). *Ae. albopictus* has been shown to be competent to transmit the same viruses as *aegypti*.

Except for its ability to diapause, *albopictus* is very similar to *aegypti* in terms of breeding sites, host choice, and ease of rearing in the lab. In fact, when *albopictus* invades territory occupied by *aegypti*, it often outcompetes *aegypti*, at least in parts of its habitat (chapter 7). Generally, in towns or cities where both *aegypti* and *albopictus* are found, the former is more common in densely urbanized localities, while the latter is more common in green areas, like parks and gardens (Swan et al. 2018). The history of the spread of *albopictus* mirrors that of *aegypti*, except that Asia is its native home, and its spread began 400 years later, in the mid-20th century. Unmounted automobile tires are a likely primary conduit for its dispersal (e.g., Hawley et al. 1985). The automobile tire industry shifted to Asia shortly before *albopictus* escaped (Kennedy and Lucks 1999). The commercial ornamental plant trade is another documented mode of transport for this species (e.g., Linthicum et al. 2003).

A complete genome assembly for *albopictus* is available (Palatini et al. 2020), and considerable work on population genetics has been done, especially in tracing the history of introductions (Beebe et al.

2013; Kotsakiozi et al. 2017a; Manni et al. 2017; Motoki et al. 2019; Pichler et al. 2019; Gloria-Soria et al. 2022b; Vavassori et al. 2022). A chikungunya outbreak has been attributed to *albopictus*, which took place on Réunion Island in 2004–2007 (Delatte et al. 2008). Chikungunya in Italy was also due to *albopictus* (Caputo et al. 2020). The physiology and genetics of its ability to diapause has also been studied (Urbanski et al. 2010, 2012; Armbruster 2016; Boyle et al. 2021).

Anopheles

The majority (96%) of deaths from malaria occur in sub-Saharan Africa, with most due to *Plasmodium falciparum* and *gambiae* s.l. Thus there's justification for featuring this mosquito. Africa does have other anophelines that transmit malaria regularly, with *funestus* second only to *gambiae*. The distribution of *funestus* s.l. coincides with *gambiae* s.l., and, like *gambiae*, it is composed of a number of cryptic taxa (Coetzee and Koekemoer 2013; Dia et al. 2013). The evolutionary genetics of two of the cryptic species of *funestus* reveal a history very closely paralleling *gambiae* and *coluzzii* (chapter 10). Inversion polymorphism fosters its spread to a new environment (e.g., rice fields), with divergence elsewhere in the genome providing reproductive isolation, protecting the new adaptations (Small et al. 2023).

An. moucheti and *nili*, ranking just behind *gambiae* and *funestus*, are important African vectors of malaria. They have narrower distributions, and their ecology overlaps with with the first two, but they also have some unique traits (Antonio-Nkondijio and Simard 2013). Sinka et al. (2010) provide additional data on African malaria vectors.

In South America, *darlingi* is the major malaria vector. Charlwood (1996) and Hiwat and Bretas (2011) reviewed the behavior and ecology of this vector. A complete genome has been obtained (Marinotti et al. 2013). Like many other anopheline malaria vectors, *darlingi* is not easy to rear in the lab, although at least one colony has been established (Puchot et al. 2022).

The complexity and number of malaria vectors in Asia are greater than in Africa and the Americas. Sinka et al. (2011) list 19 malaria vectors in Asia and the South Pacific. In Southeast Asia, *dirus* is the most important, while in India, *stephensi*, *fluviatilis*, and *culicifaies* are the dominant malaria vectors.

An. stephensi is of interest for two other reasons. This urban-adapted malaria vector has recently invaded Africa, raising the possibility of urban malaria affecting this continent, which had been largely free of malaria in large cities (Faulde et al. 2014). Second, *stephensi* is one of the easiest anophelines to rear in the laboratory, so it has been the subject of much (if not most) lab-based research on anophelines.

While these tropical regions in Africa, South America, and Asia suffer the most from malaria, other countries and parts of the temperate world have anophelines that are capable of transmitting malaria and have done so in the past, such as *freeborni* in California. Europe also has several potential malaria vectors (Bertola et al. 2022).

Ecology and Behavior

The choices of the "other" mosquitoes highlighted above are based primarily on their ongoing and potential impacts on human health. In addition, two other mosquitoes are worth mentioning as good models for ecological and behavioral work.

Aedes triseriatus and Close Relatives

Ae. triseriatus and its three close relatives are all temperate, tree hole–breeding mosquitoes found throughout North America, especially in the eastern half of the United States. As such, they are convenient and accessible models for North American researchers wanting to study natural populations of mosquitoes. The two best-studied species are *triseriatus* and *hendersoni*, which are sympatric in much of their ranges.

Egg hatching in this group is well studied. What has been dubbed "installment hatching" has been documented. Upon flooding, only a portion of the eggs hatch, with more hatching at each subsequent flooding. Khatchikian et al. (2009) related this to rainfall, with populations in regions with more rain having less-delayed hatches, compared with populations in regions with a lower amount of rain. The implication is that this pattern of hatching is an adaptation to the predictability of sufficient water persisting, allowing larvae to develop through to adults.

Installment hatching may also prevent smaller larvae from being cannibalized by larger instars (Koenekoop and Livdahl 1986). Moreover, eggs can go into diapause in response to decreasing day lengths, in contrast to egg diapause being induced in the female laying the eggs (Shroyer and Craig 1983).

The competitive interactions of these two species are among the best studied of any temperate mosquitoes. Larval habitats are divided by height with *hendersoni* found in higher tree holes, and *triseriatus* closer to the ground. Of note, when allopatric, *hendersoni* preferentially oviposits in egg traps closer to the ground (Fitzgerald and Livdahl 2019). This is consistent with the two species niche partitioning when they are sympatric.

Larvae in this group are also often found with predacious *Toxorhynchites* larvae, which can cause considerable mortality. The behavior of *triseriatus* larvae changes in response to presence of *Toxorynchites* (Juliano and Reminger 1992). *Toxorhynchites* may also affect the outcome of competitor interactions (Bradshaw and Holzapfel 1983). Moreover, parasites play a role in competition within this group (Aliabadi and Juliano 2002).

In addition to *aegypti* being replaced or outcompeted by introduced *albopictus* (chapter 7), *triseriatus* has also been affected by this invasion (Livdahl and Wiley 1991). One study attempted to elucidate three species' competitive interactions (*aegypti*, *albopictus*, and *triseriatus*), producing complex and largely inconclusive results (Ho et al. 1989).

Medically, *triseriatus* is a vector for the La Crosse virus, a moderately significant health risk. Of most interest, this mosquito and virus were the first to be documented as displaying transovarial transmission, with the virus persisting across mosquito generations through egg cytoplasm (Watts et al. 1973). This has since been demonstrated in a number of mosquito-borne viruses, including dengue.

Wyeomyia smithii

Wyeomyia smithii larvae live in pitcher plants (*Sarracenia purpurea*). This mosquito is distributed in the northern United States, with disjunct populations in the Gulf Coast of Mexico and coastal North Carolina

(Armbruster et al. 1998). It is the only known species to have both obligate blood-requiring populations (northern) and obligate nonbiting populations (Bradshaw et al. 2018). Bradshaw and Holzapfel (2017) studied variation in this species' circadian behavior and the involvement of seasonal day length changes in inducing diapause. The role of photoperiod versus temperature has also been evaluated (Bradshaw et al. 2004).

Thus *smithii* present a unique system for examining a plethora of traits associated with mosquitoes: coevolution with a host plant, a transition from vertebrate biting to nonbiting, and environmental cues inducing diapause.

REFERENCES

Aardema, M. L. B., et al. 2020. Global evaluation of taxonomic relationships and admixture within the *Culex pipiens* complex of mosquitoes. *Parasit. Vectors* 13:8.

Abong'o, B., et al. 2021. Comparison of four outdoor mosquito trapping methods as potential replacements for human landing catches in western Kenya. *Parasit. Vectors* 14:320.

Achee, N. L., et al. 2015. Considerations for the use of human participants in vector biology research: a tool for investigators and regulators. *Vector-Borne Zoo. Dis.* 15:89–102.

Afify, A., and C. G. Galizia. 2015. Chemosensory cures for mosquito oviposition site selection. *J. Med. Entomol.* 52:120–130.

Afrane, Y. A., A. K. Githeko, and G. Yan. 2012. The ecology of *Anopheles* mosquitoes under climate change: case studies from the effects of deforestation in East African highlands. *Ann. NY Acad. Sci.* 1249:204–210.

Ageep, T., et al. 2014. Participation of irradiated *Anopheles arabiensis* males in swarms following filed release in Sudan. *Malaria J.* 13:484.

Albeny-Simoes, D., et al. 2014. Attracted to the enemy: *Aedes aegypti* prefers oviposition sites with predator-killed conspecifics. *Oecologia* 175:481–492.

Aldersley, A., and L. J. Cator. 2019. Female resistance and harmonic convergence influence male mating success in *Aedes aegypti*. *Sci. Rep.* 9:2145.

Alenou, L. D., and J. Etang. 2021. Airport malaria in non-endemic areas: new insights into mosquito vectors, case management and major challenges. *Microorganisms* 9:2160.

Aliabadi, B. W., and S. A. Juliano. 2002. Escape from gregarine parasites affects the competitive interactions of an invasive mosquito. *Biol. Invasions* 4:283–297.

Anderson, J. F., et al. 2007. Nocturnal activity of mosquitoes (Diptera: Culicidae) in a West Nile Virus focus in Connecticut. *J. Med. Entomol.* 44:1102–1108.

Andreadis, T. G., P. M. Armstrong, and W. I. Bajwa. 2010. Studies on hibernating populations of *Culex pipiens* from a West Nile virus endemic focus in New York City: parity rates and isolation of West Nile virus. *J. Am. Mosq. Contr. Assoc.* 26:257–264.

Anopheles gambiae 1000 Genomes Consortium. 2017. Genetic diversity of the African malaria vector *Anopheles gambiae*. *Nature* 552:96–100.

Anopheles gambiae 1000 Genomes Consortium. 2020. Genome variation and population structure among 1142 mosquitoes of the African malaria vector species *Anopheles gambiae* and *Anopheles coluzzii*. *Genome Res.* 30:1533–1546.

Antonio-Nkondijio, C., and F. Simard. 2013. Highlights of *Anopheles nili* and *Anopheles moucheti* malaria vectors in Africa. Pp. 221–238 in S. Manguin, ed., *Anopheles Mosquitoes: New Insights into Malaria Vectors*. Rijeka, Croatia: Intech.

Apostol, B. L., et al. 1994. Use of randomly amplified polymorphic DNA amplified by polymerase chain reaction markes to estimate the number of *Aedes aegypti* families at oviposition sites in San Juan, Puerto Rico. *Am. J. Trop. Med. Hyg.* 51:89–97.

Arensburger, P., et al. 2010. Sequencing of *Culex quinquefasciatus* establishes a platform for mosquito comparative genomics. *Science* 330:86–88.

Arich, S., et al. 2021. No association between habitat, autogeny and genetics in Moroccan *Culex pipiens* populaltions. *Parasit. Vectors* 15:405.

Armbruster, P. A. 2016. Photoperiodic diapause and establishment of *Aedes albopictus* (Diptera: Culicidae) in North America. *J. Med. Entomol.* 53:1013–1023.

Armbruster, P. A., W. E. Bradshaw, and C. M. Holzapfel. 1998. Effects of postglacial range expansion on allozyme and quantitative genetic variation of the pitcher-plant mosquito, *Wyeomyia smithii*. *Evolution* 52:1697–1704.

Aryan, A., et al. 2020. *Nix* alone is sufficient to convert female *Aedes aegypti* into fertile males and *myo-sex* is needed for male flight. *Proc. Natl. Acad. Sci. USA* 117:17702–17709.

Assogba, L., et al. 2014. Characterization of swarming and mating behaviour between *Anopheles coluzzii* and *Anopheles melas* in a sympatry area of Benin. *Acta Trop.* 132 (Suppl.):S553–S563.

Avilla, F. W., et al. 2011. Insect seminal fluid proteins: identification and function. *Ann. Rev. Entomol.* 56:21–40.

Azar, D., et al. 2023. The earliest fossil mosquito. *Curr. Biol.* 33:5240–5246.

Baber, I., et al. 2010. Population size and migration of *Anopheles gambiae* in the Bancoumana region of Mali and their significance for efficient vector control. *PLOS ONE* 5:e10270.

Baesahn, R. 2021. Swarming behavior of *Anopheles gambiae* (sensu lato). *J. Med. Entomol.* 59:56–66.

Bargielowski, I., and L. P. Lounibos. 2014. Rapid evolution of reduced receptivity to interspecific mating in the dengue vector *Aedes aegypti* in response to satyrization by invasive *Aedes albopictus*. *Evol. Ecol.* 28:193–203.

Bargielowski, I., and L. P. Lounibos. 2016. Satyrization and satyrization-resistance in competitive displacements of invasive mosquito species. *Insect Sci.* 23:162–174.

Bargielowski, I., et al. 2015. Widespread evidence for interspecific mating between

Aedes aegypti and *Aedes albopictus* (Diptera: Culicidae) in nature. *Infec. Genet. Evol.* 36:456–461.

Barr, A. R. 1957. The distribution of *Culex pipiens pipiens* and *Culex pipiens quinquefasciatus* in North America. *Am. J. Trop Med. Hyg.* 6:153–165.

Barr, A. R. 1967. Occurrence and distribution of the *Culex pipiens* complex. *Bull. World Health Organ.* 37:293–296.

Barrera, R., et al. 2008. Unusual productivity of *Aedes aegypti* in septic tanks and its implications for dengue control. *Med. Vet. Entomol.* 22:62–69.

Barron, M. G., et al. 2019. A new species in the major malaria vector complex sheds light on reticulated species evolution. *Sci. Rep.* 9:14753.

Bartiliol, B., et al. 2021. Bionomics and ecology of *Anopheles merus* along the East and Southern Africa coast. *Parasit. Vectors* 14:84.

Bayoh, M. N., et al. 2014. Persistently high estimates of late night, indoor exposure to malaria vectors despite high coverage of insecticide treated nets. *Parasit. Vectors* 7:380.

Beaty, B. J., R. B. Tesh, and T. H. Aitken. 1980. Transovarial transmission of yellow fever virus in *Stegomyia* mosquitoes. *Am. J. Trop. Med. Hyg.* 29:125–132.

Beaumont, M. A., W. Zhang, and D. J. Balding. 2002. Approximate Bayesian computation in population genetics. *Genetics* 164:2025–2035.

Becker, N., A. Jost, and T. Weitzel. 2012. The *Culex pipiens* complex in Europe. *J. Am. Mosq. Contr. Assoc.* 28:53–67.

Beebe, N. W., et al. 2013. Tracing the tiger: population genetics provides valuable insights into the *Aedes* (*Stegomyia*) *albopictus* invasion of the Australasian region. *PLOS Negl. Trop. Dis.* 7:e2361.

Beebe, N. W., et al. 2021. Releasing incompatible males drives strong suppression across populations of wild and *Wolbachia*-carrying *Aedes aegypti* in Australia. *Proc. Natl. Acad. Sci. USA* 41:e2106828118.

Begun, D. J., and C. F. Aquadro. 1992. Levels of naturally occurring DNA polymorphism correlate with recombination rate in *D. melanogaster*. *Nature* 356: 519–520.

Bellow, J. E., and R. T. Carde. 2022. Compounds from human odor induce attraction and landing in female yellow fever mosquitoes (*Aedes aegypti*). *Sci. Rep.* 12:15638.

Benedict, M. Q., et al. 2018. Guidance for evaluating the safety of experimental releases of mosquitoes, emphsizing mark-release-recapture techniques. *Vector-Borne Zoo. Dis.* 18:39–48.

Benelli, G. 2018. Mating behavior of the West Nile virus vector *Culex pipiens*—role of behavioral symmetries. *Acta Trop.* 179:88–95.

Bentley, M. D., and J. F. Day. 1989. Chemical ecology and behvioral aspects of mosquito oviposition. *Ann. Rev. Entomol.* 34:401–421.

Benzon, G. L., and C. S. Apperson. 1988. Reexamination of chemically mediated oviposition behavior in *Aedes aegypti*. *J. Med. Entomol.* 25:158–164.

Bernhardt, S. A., et al. 2009. Evidence of multiple chromosomal inversions in *Aedes aegypti formosus* from Senegal. *Insect Mol. Biol.* 18:557–569.

Bertola, M., et al. 2022. Updated occurrence and bionomics of potential malaria vectors in Europe: a systematic review (2000–2021). *Parsit. Vectors* 15:88.

Besansky, N. J., and J. R. Powell. 1992. Reassociation kinetics of *Anopheles gambiae* (Diptera: Culicidae) DNA. *J. Med. Entomol.* 29:125–128.

Besansky, N. J., C. A. Hill, and C. Costantini. 2004. No accounting for taste: host preference in malaria vectors. *Trends Parasitol.* 20:249–251.

Besansky, N. J., et al. 2003. Semipermeable species boundaries between *Anopheles gambiae* and *Anopheles arabiensis*: evidence from multilocus DNA sequence variation. *Proc. Natl. Acad. Sci. USA* 100:10818–10823.

Bhattacharjee, S., et al. 2023. Mechanism of immune evasion in mosquito-borne diseases. *Pathogens* 12:635.

Bialosuknia, S. M., et al. 2022. Adaptive evolution of West Nile virus facilitated increased transmissibility and prevalence in New York State. *Emerg. Microbes Infect.* 11:2056521.

Biedler, J. K., et al. 2023. On the origin and evolution of the mosquito male-determining factor *Nix*. *Mol. Biol. Evol.* 41. https://doi.org/10.1093/molbev/msad276.

Black, W. C., and S. A. Bernhardt. 2009. Abundant nuclear copies of mitochondrial origin (NUMTs) in the *Aedes aegypti* genome. *Insect Mol. Biol.* 18:705–713.

Blaustein, J., A. Sadeh, and L. Blaustein. 2014. Influence of fire salamander laravae on among-pool distribution of mosquito egg rafts: oviposition habitat selection or egg raft predation? *Hydrobiologia* 723:157–165.

Bockarie, M. J., et al. 2009. Role of vector control in the global program to eliminate lymphatic filariasis. *Ann. Rev. Entomol.* 54:469–487.

Bogh, C., et al. 2003. Localized breeding of the *Anopheles gambiae* complex (Diptera: Culicidae) along the River Gambia, West Africa. *Bull. Entomol. Res.* 93:279–287.

Bonds, J. A. S., C. M. Collins, and L.-C. Gouagna. 2022. Could species-focused suppression of *Aedes aegypti*, the yellow fever mosquito, and *Aedes albopictus*, the tiger mosquito, affect interacting predators?: an evidence synthesis from the literature. *Pest Manag. Sci.* 78:2729–2745.

Borkent, A. 2008. The frog-biting midges of the world (Corethrellidae: Diptera). *Zootaxa* 1804:1–456.

Borkent, A., and D. A Grimaldi. 2004. The earliest fossil mosquito (Diptera: Culicidae), in mid-Cretaceous amber. *Ann. Entomol. Soc. Am.* 97:882–888.

Bouckaert, R., et al. 2014. BEAST 2: a software platform for Bayesian evolutionary analysis. *PLOS Comput. Biol.* 10: e1003537.

Bowen, M. F. 1992. Patterns of sugar feeding in diapausing and nondiapausing *Culex pipiens* (Diptera: Culicidae) females. *J. Med. Entomol.* 29:843–849.

Boyle, J. H., et al. 2021. A linkage-based genome assembly for the mosquito *Aedes albopictus* and identification of chromosomal reigons affecting diapause. *Insects* 12:020167.

Bradshaw, W. E., and C. M. Holzapfel. 1983. Predator-mediated, non-equilibrium coexistence of tree-hole mosquitoes in southeastern North America. *Oecologia* 57:239–256.

Bradshaw, W. E., and C. M. Holzapfel. 2017. Natural variation and genetics of photoperiodism in *Wyeomyia smithii*. *Adv. Genet.* 99:39–71.

Bradshaw, W. E., P. A. Zani, and C. M. Holzapfel. 2004. Adaptation to temperate climates. *Evolution* 58:1748–1762.

Bradshaw, W. E., et al. 2017. Evolutionary transition from blood feeding to obligate nonbiting in a mosquito. *Proc. Natl. Acad. Sci. USA* 115:1009–1014.

Brady, O. J., et al. 2013. Modelling adult *Aedes aegypti* and *Aedes albopictus* survival at different temperatures in laboratory and field settings. *Parasit. Vectors* 6:351.

Braks, M. A. H., et al. 2004. Interspecific competition between two invasive species of container mosquitoes, *Aedes aegypti* and *Aedes albopictus* (Diptera: Culicidae), in Brazil. *Ann. Entomol. Soc. Am.* 97:130–139.

Briegel, H. 1990. Fecundity, metabolism, and body size in *Anopheles* (Diptera: Culicidae), vectors of malaria. *J. Med. Entomol.* 27:839–850.

Brown, J. E., et al. 2011. *Aedes aegypti* mosquitoes imported into the Netherlands. *Emerg. Infect. Dis.* 17:2335–2337.

Brown, J. E., et al. 2014. Human impacts have shaped historical and recent evolution in *Aedes aegypti*, the yellow fever mosquito. *Evolution* 68:514–535.

Brugueras, S., et al. 2020. Environmental drivers, climate change and emergent diseases transmitted by mosquitoes and their vectors in southern Europe. *Environ. Res.* 191:110038.

Bruno, D. W., and B. R. Laurence. 1979. The influence of the apical droplet of *Culex* egg rafts on oviposition of *Culex pipiens fatigans* (Diptera: Culicidae). *J. Med. Entomol.* 16:300–305.

Bryant, J. H., and F. Gebert. 1976. Indentification of members of the *Anopheles gambiae* complex from Mauritius. *Trans. Roy. Soc. Trop. Med. Hyg.* 70:339.

Bryant, J. H., and B. A. Southgate. 1978. Studies of forced mating techniques on Anopheline mosquitoes. *Mosq. News* 38:338–342.

Buffington, J. D. 1972. Hibernaculum choice in *Culex pipiens*. *J. Med. Entomol.* 9:128–132.

Bullini, L., M. Coluzzi, and A. P. Bianchi Bullini. 1976. Biochemical variants in the study of multiple insemination in *Culex pipiens* L. (Diptera, Culicidae). *Bull. Entomol. Res.* 65:683–685.

Buonaccorsi, J. P., L. C. Harrington, and J. D. Edman. 2003. Estimation and comparison of mosquito survival rates with release-recapture-removal data. *J. Med. Entomol.* 40:6–17.

Calhoun, L. M., et al. 2007. Combined sewage overflows (CSO) are major urban breeding sites for *Culex quinquefasciatus* in Atlanta, Georgia. *Am. J. Trop. Med. Hyg.* 77:478–484.

Calvez, E., et al. 2016. Genetic diversity and phylogeny of *Aedes aegypti*, the main arbovirus vector in the Pacific. *PLOS Negl. Trop. Dis.* 10: e0004374.

Cannon, W. F. 1961. The impact of uniformitarianism. *Proc. Am. Phil. Soc.* 105:301–314.

Cansado-Utrilla, C., et al. 2021. The microbiome of mosquito vectorial capacity: rich potential for discovery and translation. *Microbiome* 9:111.

Capone, A., et al. 2013. Interactions between *Asaia*, *Plasmodium* and *Anopheles*: new insights into mosquito symbiosis and implications for malaria symbiotic control. *Parasit. Vectors* 6:182.

Caputo, B., et al. 2007. Comparative analysis of epicuticular lipid profiles of sympatric and allopatric field populations of *Anopheles gambiae* s.s. molecular forms and *An. arabiensis* from Burkina Faso (West Africa). *Insect Biochem. Mol. Biol.* 37:389–398.

Caputo, B., et al. 2020. A comparative analysis of the 2007 and 2017 Italian chikungunya outbreaks and implicaitons for public health response. *PLOS Negl. Trop. Dis.* 14:e0008159.

Caputo, B., et al. 2024. Populationn genomic evidence of a putative “far-west” African cryptic taxon in the *Anopheles gambiae* complex. *Commun. Biol.* 7:1115.

Caragata, E. P., C. V. Tikhe, and G. Dimopoulos. 2019. Curious entaglements: interactions between mosquitoes, their microbiota, and arboviruses. *Curr. Opin. Virol.* 37:26–36.

Carlson, C. J., et al. 2023. Rapid range shifts in African *Anopheles* mosquitoes over the last century. *Biol. Lett.* 19:20220365.

Carrieri, M., et al. 2003. On the competition occurring between *Aedes albopictus* and *Culex pipiens* (Dipera: Culicidae) in Italy. *Environ. Entomol.* 32:1313–1321.

Carson, H. L. 1959. Conditions which promote or retard the formation of species. *Cold Spr. Harb. Symp. Quant. Biol.* 24:87–105.

Carvajal, T. M., et al. 2020. Fine-scale population genetic structure of dengue mosquito vector, *Aedes aegypti*, in metropolitan Manila, Philippines. *PLOS Negl. Trop. Dis.* 14:e0008279.

Cator, L. J., et al. 2011. Behavioral observations of sound recordings of free-flight mating swarms of *Ae. aegypti* (Diptera: Culicidae) in Thailand. *J. Med. Entomol.* 48:941–946.

Chadee, D. D. 1984. *Aedes aegypti* aboard boats at Port-of-Spain, Trinidad, West Indies (1972–1982). *Mosq. News* 44:1–3.
Chadee, D. D., and R. Martinez. 2016. *Aedes aegypti* (L.) in Latin American and Caribbean region: with growing evidence for vector adaptation to climate change? *Acta Trop.* 156:137–143.
Chadee, D. D., J. C. Beier, and R. T. Mohammed. 2002. Fast and slow blood-feeding durations of *Aedes aegypti* mosquitoes in Trinidad. *J. Vector Ecol.* 27:172–177.
Chadee, D. D., J. M. Sutherland, and J. R. L. Gilles. 2014. Diel sugar feeding and reproductive behaviours of *Aedes aegypti* in Trinidad: with implications for mass release of sterile mosquitoes. *Acta Trop.* 132 (Suppl.): S86–S90.
Chadee, D. D., R. A. Ward, and R. J. Novak. 1998. Natural habitats of *Aedes aegypti* in the Caribbean—a review. *J. Am. Mosq. Contr. Assoc.* 14:5–11.
Chan, M., and M. A. Johansson. 2012. The incubation periods of dengue viruses. *PLOS ONE* 7:e50972.
Chandra, G., et al. 2008. Mosquito control by larvivorous fish. *Indian J. Med. Res.* 127:13–27.
Chandrasegaran, K., et al. 2018. Playing it safe?: behavioural responses of mosquito larvae encountering a fish predator. *Ethol. Ecol. Evol.* 30:70–87.
Charlwood, J. D. 1996. Biological variation in *Anopheles darlingi* Root. *Mem. Inst. Oswaldo Cruz* 91:391–398.
Charlwood, J. D. 2020. *The Ecology of Malaria Vectors*. Boca Raton, FL: CRC Press.
Charlwood, J. D., et al. 2002. The warming and mating behavior of *Anopheles gambiae* s.s. (Diptera: Culicidae) from São Tomé Island. *J. Vector Ecol.* 27:178–183.
Chaves, L. F., et al. 2009. Combined sewage overflow enhances oviposition of *Culex quinquefasciatus* (Dipetera: Culicidae) in urban areas. *J. Med. Entomol.* 46:220–226.
Chen, H., U. Filliner, and G. Yan. 2006. Oviposition behavior of female *Anopheles gambiae* in western Kenya inferred from microsatellite markers. *Am. J. Trop. Med. Hyg.* 75:246–250.
Cheng, C., et al. 2012. Ecological genomics of *Anopheles gambiae* along a latitutidinal cline: a population-resequencing approach. *Genetics* 190:1417–1432.
Cheng, C., et al. 2018. Systems genetic analysis of inversion polymorhisms in the malaria mosquito *Anopheles gambiae*. *Proc. Natl. Acad. Sci. USA* 115:E7005–E7014.
Chevillon, C., et al. 1998. Migration/selection balance and ecotypic differentiation in the mosquito *Culex pipiens*. *Mol. Ecol.* 7:197–208.
Chouaia, B., et al. 2010. Molecular evidence for multiple infections as revealed by typing of *Asaia* bacterial symbionts of four mosquito species. *App. Environ. Microbiol.* 76:7444– 7450.

Christie, M. 1958. Predation on larvae of *Anopheles gambiae* Giles. *J. Trop. Med. Hyg.* 61:168–176.

Christophers, R. 1960. *Aedes aegypti: The Yellow Fever Mosquito*. Cambridge: Cambridge University Press.

Ciota, A. T., et al. 2011. Emergence of *Culex pipiens* from overwintering hibernacula. *J. Am. Mosq. Contr. Assoc.* 27:21–29.

Clements, A. N. 1992. *The Biology of Mosquitoes*. Vol. 1, *Development, Nutrition, and Reproduction*. London: Chapman & Hall.

Clements, A. N. 1999. *The Biology of Mosquitoes*. Vol. 2, *Sensory Reception and Behaviour*. Wallingford, UK: CABI.

Clements, A. N. 2012. *The Biology of Mosquitoes*. Vol. 3, *Viral and Bacterial Symbionts*. Cambridge: Cambridge University Press.

Coetzee, M., and L. L. Koekemoer. 2013. Molecular systematics and insecticide resistance in the major African malaria vector *Anopheles funestus*. *Ann. Rev. Entomol.* 58:393–412.

Coetzee, M., M. Craig, and D. le Suer. 2000. Distribution of African malaria mosquitoes belonging to the *Anopheles gambiae* complex. *Parasitol. Today* 16:74–77.

Coetzee, M., et al. 2013. *Anopheles coluzzii* and *Anopheles amharicus*, new members of the *Anopheles gambiae* complex. *Zootaxa* 3619:246–274.

Collins, C. M., et al. 2019. Effects of the removal or reduction in density of the malaria mosquito, *Anopheles gambiae* s.l., on interacting predators and competitors in local ecosystems. *Med. Vet. Entomol.* 33:1–15.

Colton, Y. M., D. D. Chadee, and D. W. Severson. 2003. Natural skip oviposition of the mosquito *Aedes aegypti* indicated by codominant genetic markers. *Med. Vet. Entomol.* 17:195–204.

Coluzzi, M. 1964. Morphological divergences in the *Anopheles gambiae* complex. *Riv. Malariol.* 43:197–232.

Coluzzi, M. 1982. Spatial distribution of chromosomal inversions and speciation in anopheline mosquitoes. Pp. 143–153 in C. Barigozzi, ed. *Mechanisms of Speciation*. New York: Alan R. Liss.

Coluzzi M, V. Petrarca, and M. A. DiDeco. 1985. Chromosomal inversion intergradation and incipient speciation in *Anopheles gambiae*. *Bull. Zool.* 52:45–63.

Coluzzi M., et al. 1979. Chromosomal differentiation and adaptation to human environments in the *Anopheles gambiae* complex. *Trans. R. Soc. Trop. Med. Hyg.* 73:483–497.

Coluzzi, M., et al. 2002. A polytene chromosome analysis of the *Anopheles gambiae* species complex. *Science* 298:1415–1418.

Conway, G. R., M. Trpis, and G. A. H. McClelland. 1974. Population parameters of the mosquito *Aedes aegypti* (L.) estimated by mark-release-recapture in a suburban habitat in Tanzania. *J. Animal Ecol.* 43:289–304.

Coon, K. L., M. R. Brown, and M. R. Strand. 2016. Mosquitoes host communities of bacteria that are essential for development but vary greatly between local habitats. *Mol. Ecol.* 25:5806–5826.

Cornel, A. J., et al. 2003. Differences in extent of genetic introgression between sympatric *Culex pipiens* and *Culex quniquefasciatus* (Diptera: Culicidae) in California and South Africa. *J. Med. Entomol.* 40:36–51.

Cornet, S., et al. 2019. Avian malaria alters the dynamics of blood feeding in *Culex pipiens* mosqitoes. *Malar. J.* 18:82.

Cosme, L. V., et al. 2024. A genotyping array for the globally invasive vector mosquito, *Aedes albopictus*. *Parasit. Vect.* 17:106.

Costantini, C., et al. 1993. A new odour-baited trap to collect host-seeking mosquitoes. *Parassitologia* 35:5–9.

Costantini, C., et al. 1996. Density, survival and dispersal of *Anopheles gambiae* complex mosquitoes in a West African Sudan savanna village. *Med. Vet. Entomol.* 10:203–219.

Costantini, C., et al. 1998. Odor-mediated host preferences of West African mosquitoes, with particular reference to malaria vectors. *Am. J. Trop. Med. Hyg.* 58:56–63.

Costantini, C., et al. 2009. Living at the edge: biogeographic patterns of habitat segregation conform to speciation by niche expansion in *Anopheles gambiae*. *BMC Ecol.* 9:16.

Costero, A., et al. 1999. Survival of starved *Aedes aegypti* (Diptera: Culicidae) in Puerto Rico and Thailand. *J. Med. Entomol.* 36:272–276.

Couper, L. I., et al. 2021. How will mosquitoes adapt to climate warming? *eLife* 10:e69630.

Couret, J., and M. Q. Benedict. 2014. A meta-analysis of the factors influencing development rate variation in *Aedes aegypti* (Diptera: Culicidae). *BMC Ecol.* 14:3.

Coutinho-Abreu, I. V., J. A. Riffell, and O. S. Akbari. 2022. Human attractive cues and mosquito host-seeking behavior. *Trends Parasitol.* 38:246–262.

Craig, G. B. 1967. Mosquitoes: female monogamy induced by male accessory gland substance. *Science* 156:1499–1501.

Crawford, J., et al. 2015. Reticulate speciation and barriers to introgression in the *Anopheles gambiae* species complex. *Genome Biol. Evol.* 7:3116–3131.

Crawford, J., et al. 2017. Population genomics reveals that an anthropophilic population of *Aedes aegypti* mosquitoes in West Africa recently gave rise to American and Asian populations of this major disease vector. *BMC Biol.* 15:16.

Crawford, J., et al. 2020. Efficient production of male *Wolbachia*-infected *Aedes aegypti* mosquitoes enables large-scale suppresion of wild populations. *Nat. Biotechnol.* 38:482–492.

Crawford, J. E., et al. 2024. Sequencing 1206 genomes reveals origin and movement

of *Aedes aegypti* driving increased dengue risk. *bioRχiv* 7.23.604830. https://doi.org/10.1101/2024.07.23.604830.
Crosby, M. C. 2006. *The American Plague*. New York: Berkeley Books.
Cui, F., et al. 2007. Genetic differentiation of *Culex pipiens* (Diptera: Culicidae) in China. *Bull. Entomol. Res.* 97:291–297.
Dabire, K. R., et al. 2013. Assortative mating in mixed swarms of the mosquito *Anopheles gambiae* s.s. M and S molecular forms, in Burkina Faso, West Africa. *Med. Vet. Entomol.* 27:298–312.
Dabire, K. R., et al. 2014. Occurrence of natural *Anopheles arabiansis* swarms in an urban area of Bobo-Dialasso City, Burkina Faso, West Africa. *Acta Trop.* 132:S35–S41.
Da Costa–Ribeiro, M. C. V., R. Lourenço-de-Oliveria, and A.-B. Failloux. 2016. Geographical and temporal genetic patterns of *Aedes aegypti* populations in Rio de Janeiro, Brazil. *Trop. Med. Intnat. Health* 11:1276–1285.
Dada, N., et al. 2018. Whole metagenome sequencing reveals links between mosquito microbiota and insecticide resistance in malaria vectors. *Sci. Rep.* 8:2084.
Dada, N., et al. 2021. Considerations for mosquito microbiome research from the Mosquito Microbiome Consortium. *Microbiome* 9:36.
Dambach, P. 2020. The use of aquatic predators for larval control of mosquito disease vectors: opportunities and limitations. *Biol. Contr.* 150:104357.
Damiani, C., et al. 2010. Mosquito-bacteria symbiosis: the case of *Anopheles gambiae* and *Asaia*. *Microb. Ecol.* 60:644–654.
Das, S., L. Garver, and G. Dimopoulos. 2007. Protocol for mosquito rearing (*A. gambiae*). *JoVE Journal: Biology* 5. https://app.jove.com/t/221.
Davis, N. C. 1932. The effect of various temperatures in modifying the extrinsic incubation period of the yellow fever virus in *Aedes aegypti*. *Am. J. Epidemiol.* 16:163–176.
Day, J. F. 2016. Mosquito oviposition behavior and vector control. *Insects* 7:65.
Delatte, H., et al. 2008. *Aedes albopictus*, vector of chikungunya and dengue viruses in Réunion Island: biology and control. *Parasites* 15:3–13.
Della Torre, A., Z. Tu, and V. Petrarca. 2005. On the distribution and genetic differentiation of *Anopheles gambiae* s.s. molecular forms. *Insect Biochem. Mol. Biol.* 35:755–769.
Denlinger, D. L., and P. A. Armbruster. 2014. Mosquito diapause. *Ann. Rev. Entomol.* 59:73–93.
Dennison, N. J., N. Jupatanakul, and G. Dimopoulos. 2014. The mosquito microbiota influences vector competence for human pathogens. *Curr. Opin. Insect Sci.* 3:6–13.
Dexter, J. S. 1913. Mosquitoes pollinating orchids. *Science* 37:867.
Dhileepan, K. 1997. Physical factors and chemical cues in the oviposition behavior

of the arboviral vectors *Culex annulirostris* and *Culex molestus* (Diptera: Culicidae). *Environ. Entomol.* 26:318–326.
Dia, I., M. W. Guelbeogo, and D. Ayala. 2013. Advances and perspectives in the study of the malaria mosquito *Anopheles funestus*. Pp. 197–220 in S. Manguin, ed., *Anopheles Mosquitoes: New Insights into Malaria Vectors*. Rijeka, Croatia: Intech.
Diabate, A., et al. 2005. Development of the molecular forms of *Anopheles gambiae* (Diptera: Culicidae) in different habitats: a transplant experiment. *J. Med. Entomol.* 42:548–553.
Diabate, A., et al. 2007. Evaluating the effect of postmating isolation between molecular forms of *Anopheles gambiae* (Dipera: Culicidae). *J. Med. Entomol.* 44:60–64.
Diabate, A., et al. 2008. Evidence for divergent selection between the molecular forms of *Anopheles gambiae*: role of predation. *BMC Evol. Biol.* 8:5.
Diabate, A., et al. 2011. Spatial distribution and male mating success of *Anopheles gambiae* swarms. *BMC Evol. Biol.* 11:184.
Dickens, B. L, and H. L. Brant. 2014. Effects of marking methods and fluorescent dusts on *Aedes aegypti* survival. *Parsit. Vectors* 7:65.
Dickens, B. L., et al. 2018. Determining environmental and anthropogenic factors which explain the global distribution of *Aedes aegypti* and *Aedes albopictus*. *BMJ Glob. Health* 3:e000801.
Dickson, L. B., et al. 2016. Reproductive incompatibility involving Senegalese *Aedes aegypti* (L.) is associated with chromosome rearrangements. *PLOS Negl. Trop. Dis.* 10: e0004626.
Dimopoulos, G., et al. 1996. Integrated genetic map of *Anopheles gambiae*: use of RAPD polymorphisms for genetic, cytogenetic STS landmarks. *Genetics* 143:953–960.
Dobzhansky, T. 1935. A critique of the species concept in biology. *Philos. Sci.* 2:344–355.
Dong, S., et al. 2022. Mosquito transgenesis for malaria control. *Trends Parasitol.* 38:54–66.
Donnelly, M. J., M. C. Licht, and T. Lehmann. 2001. Evidence for recent population expansion in the evolutionary history of the malaria vectors *Anopheles arabiensis* and *Anopheles gambiae*. *Mol. Biol. Evol.* 18:1353–1364.
Downes, J. A. 1969. The swarming and mating flight of Diptera. *Ann. Rev. Entomol.* 14:271–298.
Dumas, E., et al. 2016. Molecular data reveal a cryptic species within the *Culex pipiens* mosquito complex. *Insect Mol. Biol.* 25:800–809.
Dunn, L. C. 1965. Chapter 20 in *A Short History of Genetics*. New York: McGraw-Hill.
DuRant, S. E., and W. A. Hopkins. 2008. Amphibian predation on larval mosquitoes. *Can. J. Zool.* 86:1159–1164.

Edelman, N. B., and J. Mallet. 2021. Prevalence and adaptive impact of introgression. *Ann. Rev. Genet.* 55:265–283.

Edillo, F. E., et al. 2006. Water quality and immatures of the M and S forms of *Anopheles gambiae* s.s. and *An. arabiensis* in a Malian village. *Malar. J.* 5:35.

Edman, J. D., et al. 1998. *Aedes aegypti* (Diptera: Culicidae) movement influenced by availability of oviposition sites. *J. Med. Entomol.* 35:578–583.

Ellis, B. R., and A. D. T. Barrett. 2008. The enigma of yellow fever in East Africa. *Rev. Med. Virol.* 18:331–346.

Eltis, D., and D. Richardson. 2010. *The Transatlantic Slave Trade*. New Haven, CT: Yale University Press.

Endersby, N. M., et al. 2009. Genetic structure of *Aedes aegypti* in Australia and Vietnam revealed by microsatellite and exon primed intron crossing markers suggests feasibility of local control options. *J. Med. Entomol.* 46:1074–1083.

Endersby, N. M., et al. 2011. Changes in the genetic structure of *Aedes aegypti* (Diptera: Culicidae) populations in Queensland, Australia, across two seasons: implications for potential mosquito releases. *J. Med. Entomol.* 48:999–1007.

Endler, J. A. 1977. *Geographic Variation, Speciation, and Clines*. Princeton, NJ: Princeton University Press.

Epopa, P. S., et al. 2017. The use of sequential mark-release-recapture experiments to estimate population size, survival and dispersal of male mosquitoes of the *Anopheles gambiae* complex in Bana, a west African humid savannah village. *Parasit. Vectors* 10:376.

Erdelyan, C. N. G., et al. 2012. Functional validation of the carbon dioxide receptor genes in *Aedes aegypti* mosquitoes using RNA interference. *Insect Mol. Biol.* 21:119–127.

Eritja, R., and C. Chevillon. 1999. Interruption of chemical control and evolution of insecticide resistance genes in *Culex pipiens*. *J. Med. Entomol.* 36:41–49.

Estrada-Franco, J. G., et al. 2020. Vertebrate-*Aedes aegypti* and *Culex quinquefasciatus* (Diptera)-arbovirus transmission networks: non-human feeding revealed by meta-barcoding and next-generation sequencing. *PLOS Negl. Trop. Dis.* 14: e0008867.

Evanno, G., S. Ragnaut, and J. Goudet. 2005. Detecting the number of clusters of individuals using the software STRUCTURE: a simulation study. *Mol. Biol. Evol.* 14:2611–2620.

Evans, B. R., et al. 2015. A multipurpose high-throughput SNP chip for the dengue and yellow fever mosquito, *Aedes aegypti*. *G3* 5:711–718.

Excoffier, L., et al. 2013. Robust demographic inference from genomic and SNP data. *PLOS Genet.* 13:e1003905.

Facchinelli, L., A. Badolo, and P. J. McCall. 2023. Biology and behaviour of *Aedes aegypti* in the human environment: opportunities for vector control of arbovirus transmission. *Viruses* 15:636.

Faiman, R., et al. 2022. Isotopic evidence that aestivation allows malaria mosquitoes to persist through the dry season in the Sahel. *Nat. Ecol. Evol.* 6:1687–1699.

Farajollahi, A., et al. 2011. "Bird biting" mosquitoes and human disease: a review of the role of *Culex pipiens* complex mosquitoes in epidemiology. *Infect. Genet. Evol.* 11:1577–1585.

Faulde, M. K., L. M. Rueda, and B. A. Khaireh. 2014. First record of the Asian malaria vector *Anopheles stephensi* and its possible role in the resurgence of malaria in Djibouti, Horn of Africa. *Acta Trop.* 139:39–43.

Favia, G., et al. 1997. Molecular identification of sympatric chromosomal forms of *Anopheles gambiae* and further evidence of their reproductive isolation. *Insect Mol. Biol.* 6:377–383.

Favia, G., et al. 2007. Bacteria of the genus *Asaia* stably associate with *Anopheles stephensi*, an Asian malarial mosquito vector. *Proc. Natl. Acad. Sci. USA* 104: 9047–9051.

Fawaz, E. Y., et al. 2014. Swarming mechanisms in the yellow fever mosquito: aggregation pheromones are involved in the mating behavior of *Aedes aegypti*. *J. Vector Ecol.* 39:347–354.

Ferguson, H., and A. F. Read. 2002. Why is the effect of malaria parasites on mosquito survival still unresolved? *Trends Parasitol.* 18:256–261.

Field, E. N., et al. 2022. Semi-field and surveillance data define the natural diapause timeline for *Culex pipiens* across the United States. *Commun. Biol.* 5:1300.

Field, L. M., et al. 1999. Analysis of genetic variability in *Anopheles arabiensis* and *Anopheles gambiae* using microsatellite loci. *Insect Mol. Biol.* 8:287–297.

Fikrig, K., and L. C. Harrington. 2021. Understanding and interpreting mosquito blood feeding studies: the case of *Aedes albopictus*. *Trends Parasitol.* 37:959–975.

Fisher, R. A. 1947.The theory of linkage in polysomic inheritance. *Philos. Trans. R. Soc. Lond., B, Biol. Sci.* 233:55–87.

Fitzgerald, J., and T. Livdahl. 2019. Vertical habitat stratification in sympatric and allopatric populations of *Aedes hendersoni* and *Aedes triseriatus* (Diptera: Culicidae). *J. Med. Entomol.* 56:311–319.

Focks, D. A. 2003. A review of entomological sampling methods and indicators for dengue vectors. Reference no. TDR/IDE/Den/03.1. Geneva: UNICEF/UNDP/World Bank/WHO Special Programme for Research and Training in Tropical Diseases (TDR).

Focks, D. A., et al. 2000. Transmission thresholds for dengue in terms of *Aedes aegypti* pupae per person with discussion of their utility in source reduction efforts. *Am . J. Trop. Med. Hyg.* 62:11–18.

Fonseca, D. M., et al. 2004. Emerging vectors in the *Culex pipiens* complex. *Science* 303:1535–1538.
Fonseca, D. M., et al. 2006. Pathways of expansion and multiple introductions illustrated by large genetic differentiation among worldwide populations of the southern house mosquito. *Am. J. Trop. Med. Hyg.* 74:284–289.
Fontaine, M. C., et al. 2015. Extensive introgression in a malaria vector species complex revealed by phylogenomics. *Science* 347:1258524.
Fontenille, D., and J. R. Powell. 2020. From anonymous to public enemy: how does a mosquito become a feared arbovirus vector? *Pathogens* 9:265.
Fonzi, E., et al. 2015. Human-mediated marine dispersal influences the population structure of *Aedes aegypti* in the Philippine archipelago. *PLOS Negl. Trop. Dis.* 9:e0003829.
Foster, K. R., M. F. Jenkins, and A. C. Toogood. 1998. The Philadelphia yellow fever epidemic of 1793. *Sci. Am.* 79:88–93.
Frohne, W. C. 1964. Preliminary observations of fall swarms of *Culex pipiens* L. *Mosq. News* 24:369–376.
Fytrou, A. D., et al. 2022. Behavioural response of *Culex pipiens* biotype *moelstus* to oviposition pheromone. *J. Insect Physiol.* 138:104383.
Gao, H., et al. 2020. Mosquito microbiota and implications for disease control. *Trends Parasiol.* 36:98–111.
Garcia, G., et al. 2016. Using *Wolbachia* releases to estimate *Aedes aegypti* (Diptera: Culicidae) population size and survival. *PLOS ONE* 11:e0160196.
Gaunt, M. W., et al. 2001. Phylogenetic relationship of flaviviruses correlate with their epidemiology, disease association and biogeography. *J. Gen. Virol.* 82:1867–1876.
Gibson, G. 1985. Swarming behaviour of the mosquito *Culex pipiens quinquefasciatus*: a quantitative analysis. *Physiol. Entomol.* 10:283–296.
Giesen, C., et al. 2020. The impact of climate change on mosquito-borne diseases in Africa. *Pathog. Glob. Health* 114:287–301.
Giglioli, M. E. C. 1965. Oviposition by *Anopheles melas* and its effect on egg survival during the dry season in the Gambia, West Africa. *Ann. Entomol. Soc. Am.* 58:885–891.
Gilchrist, B. M., and J. B. S. Haldane. 1947. Sex linkage and sex determination in a mosquito, *Culex molestus*. *Hereditas* 33:175–190.
Gillespie, B. I., and P. Belton. 1980. Oviposition of *Culex pipiens* in water at different temperatures. *J. Entomol. Soc. Brit. Columbia* 77:34–36.
Gillies, M.T. 1954. The recognition of age-groups within populations of *Anopheles gambiae* by the pre-gravid rate and the sporozoite rate. *Ann. Trop. Med. Parasitol.* 48:58–74.
Gillies, M. T. 1961. Studies on the dispersion and survival of *Anopheles gambiae* Giles

in East Africa, by means of marking and release experiments. *Bull. Entomol. Res.* 52:99–127.

Gimnig, J. E., et al. 2002. Density-dependent development of *Anopheles gambiae* (Diptera: Culicidae) larvae in artificial habitats. *J. Med. Entomol.* 39:162–172.

Gimnig, J. E., et al. 2013. Incidence of malaria among mosquito collectors conducting human landing catches in Western Kenya. *Am. J. Trop. Med. Hyg.* 88:301–308.

Gloria-Soria, A., et al. 2014. Origin of the dengue fever mosquito, *Aedes aegypti*, in California. *PLOS Negl. Trop. Dis.* 8: e3029.

Gloria-Soria, A., et al. 2016a. Global genetic diversity of *Aedes aegypti*. *Mol. Ecol.* 25:5377–5395.

Gloria-Soria, A., et al. 2016b. Temporal genetic stability of worldwide *Stegomyia aegypti* (= *Aedes aegypti*) populations. *J. Med. Vet. Entomol.* 30:235–240.

Gloria-Soria, A., T. Chiodo, and J. R. Powell. 2018a. Lack of evidence for natural *Wolbachia* infections in *Aedes aegypti* (Diptera: Culicidae). *J. Med. Entomol.* 55:1354–1356.

Gloria-Soria, A., et al. 2018b. Origin of a high latitude population of *Aedes aegypti* in Washington, DC. *Am. J. Trop. Med. Hyg.* 98:445–452.

Gloria-Soria, A., et al. 2019. Genetic diversity of laboratory strains and implications for research: the case of *Aedes aegypti*. *PLOS Negl. Trop. Dis.* 13:e0007930.

Gloria-Soria, A., et al. 2022a. Origin of high latitude introductions of *Aedes aegypti* to Nebraska and Utah during 2019. *Infec. Genet. Evol.* 103:105333.

Gloria-Soria, A., et al. 2022b. Population genetics of an invasive mosquito vector, *Aedes albopictus* in Northeastern USA. *NeoBiota* 78:99–127.

Goi, J., et al. 2022. Comparison of different mosquito traps for zoonotic arbovirus vectors in Papua New Guinea. *Am. J. Trop. Med. Hyg.* 106:823–827.

Goeldi, E. A. 1905. *Os Mosquitos no Para.* Para, Brazil: Memorias do Museu Goeldi.

Gomes, B., et al. 2009. Asymmetric introgression between sympatric molestus and pipiens forms of *Culex pipiens* (Diptera: Culicidae) in the Comporta region, Portugal. *BMC Evol. Biol.* 9:262.

Gomes, B., et al. 2013. Feeding patterns of molestus and pipiens forms of *Culex pipiens* (Diptera: Culicidae) in a region of high hybridization. *Parasit. Vectors* 6:93.

Gomes, B., et al. 2015. Limited genomic divergence between intraspecific forms of *Culex pipiens* under different ecological pressures. *BMC Evol. Biol.* 15:197.

Gonzalez, P. V., P. A. Gonzalez Audino, and H. M. Masuh. 2016. Oviposition behavior in *Aedes aegypti* and *Aedes albopictus* (Diptera: Culicidae) in response to the presence of heterospecific and conspecific larvae. *J. Med. Entomol.* 53:268–272.

Gratz, N. G., R. Steffen, and W. Cocksedge. 2000. Why aircraft disinsection? *Bull. World Health Organ.* 78:995–1004.

Gray, L., et al. 2022. Back to the future: quantifying wing wear as a method to measure mosquito age. *Am. J. Trop. Med. Hyg.* 107:689–700.

Grimaldi, D., and M. S. Engels. 2005. *Evolution of the Insects*. New York: Cambridge University Press.

Guagliardo, S., et al. 2019. The genetic structure of *Aedes aegypti* populations is driven by boat traffic in the Peruvian Amazon. *PLOS Negl. Trop. Dis.* 13:e0007552.

Guegan, M., et al. 2018. The mosquito holobiont: fresh insight into mosquito-microbiota interactions. *Microbiome* 6:49.

Guerra, C. A., et al. 2014. A global assembly of adult female mosquito mark-release-recapture data to inform the control of mosquito-borne pathogens. *Parasit. Vectors* 7:276.

Gutenkunst, R. N., et al. 2009. Inferring the joint demographic history of multiple populations from multidimensional SNP frequency data. *PLOS Genet.* 5:e100695.

Gwadz, R. W. 1969. Regulation of blood meal size in the mosquito. *J. Insect Physiol.* 15:2039–2044.

Haba, Y., and L. McBride. 2022. Origin and status of *Culex pipiens* mosquito ecotypes. *Curr. Biol.* 32:R237–R248.

Haba, Y., et al. Submitted. Evolutionary history and population structure of *Culex pipiens* mosquitoes.

Hahn, M. B., et al. 2016. Reported distribution of *Aedes* (*Stegomyia*) *aegypti* and *Aedes* (*Stegomyia*) *albopictus* in the United States, 1995–2016 (Diptera: Culicidae). *J. Med. Entomol.* 53:1169–1175.

Hahn, M. B., et al. 2017. Updated reported distribution of *Aedes* (*Stegomyia*) *aegypti* and *Aedes* (*Stegomyia*) *albopictus* (Diptera: Culicidae) in the United States, 1995–2016. *J. Med. Entomol.* 54:1420–1424.

Hajkazemian, M., et al. 2021. Battleground midgut: the cost to the mosquito for hosting the malaria parasite. *Biol. Cell* 113:79–94.

Hamer, G. L., et al. 2014. Dispersal of adult *Culex* mosquitoes in an urban West Nile virus hotspot: a mark-capture study incorporating stable isotope enrichment of natural larval habitats. *PLOS Negl. Trop. Dis.* 8:e2768.

Harbach, R. E. 2007. The Culicidae (Diptera): a review of taxonomy, classification and phylogeny. *Zootaxa* 1668:591–638.

Harbach, R. E., and K. L. Knight. 1980. *Taxonomists' Glossary of Mosquito Anatomy*. Marlton, NJ: Plexus.

Harbach, R. E., and R. C. Wilkerson. 2023. The insupportable validity of mosquito subspecies (Diptera: Culicidae) and their exclusion from culicid classification. *Zootaxa* 530:1–184.

Harrington, L. C., and J. D. Edman. 2002. Indirect evidence against delayed "skip-ovipostion" behavior by *Aedes aegypti* (Diptera: Culiciade) in Thailand. *J. Med. Entomol.* 38:641–645.

Harrington, L. C., et al. 2005. Dispersal of the dengue vector *Aedes aegypti* within and between rural communites. *Am. J. Trop. Med. Hyg.* 72:209–220.

Harrington, L. C., et al. 2008. Age-dependent survival of the dengue vector *Aedes aegypti* (Diptera: Culicidae) demonstrated by simultaneous release-recapture of different age cohorts. *J. Med. Entomol.* 45:307–313.

Harris, P., D. F. Riordan, and D. Cooke. 1969. Mosquitoes feeding on insect larvae. *Science* 164:184–185.

Hartberg, W. K. 1971. Observations on the mating behaviour of *Aedes aegypti* in nature. *Bull. World Health Organ.* 45:847–850.

Hartberg, W. K., and G. B. Craig Jr. 1970. Reproductive isolation in *Stegomyia* mosquitoes, II: hybrid breakdown between *Aedes aegypti* and *A. mascarensis*. *Evolution* 24:692–703.

Hartberg, W. K., and G. A. H. McClelland. 1973. *Aedes mascarensis* MacGregor on Mauritius, II: genetic variability of field populations (Diptera: Culicidae). *J. Med. Entomol.* 10:577–582.

Hawley, W. A., et al. 1985. *Aedes albopictus* in North America: probable introduction in used tires from northern Asia. *Science* 236:1114–1116.

Heath, C., et al. 2020. Evidence of transovarial transmission of chikungunya and dengue viruses in field-caught mosquitoes in Kenya. *PLOS Negl. Trop. Dis.* 14:e0008362.

Hebert, P. D. N., and T. R. Gregory. 2005. The promise of DNA barcoding for taxonomy. *Syst. Biol.* 54:852–859.

Helinski, M. E. H., et al. 2012. Evidence of polyandry for *Aedes aegypti* in semifield enclosures. *Am. J. Trop. Med. Hyg.* 86:635–641.

Hemme, R. R., et al. 2010. Influence of urban landscapes on population dynamics in a short-distance migrant mosquito: evidence for the dengue vector *Aedes aegypti*. *PLOS Negl. Trop. Dis.* 4:e634.

Hemming-Schroeder, E., et al. 2020. Ecological drivers of genetic connectivity for African malaria vectors *Anopheles gambiae* and *An. arabiensis*. *Sci. Rep.* 10:19946.

Herrera-Varela, M., et al. 2014. Habitat discrimination by gravid *Anopheles gambiae* sensu lato—a push-pull system. *Malar. J.* 13:133.

Herschel, J. 1836. Letter to Charles Lyell, February 20.

Hickner, P. V., et al. 2013. Composite linkage map and enhanced genome map for *Culex pipiens* complex mosquitoes. *J. Hered.* 104:649–655.

Highton, R. B., and E. C. C. van Someren. 1970. Transportation of mosquitoes between international airports. *Bull. World Health Organ.* 42:127–135.

Hillyer, J. F. 2016. Insect immunology and hematopoiesis. *Dev. Comp. Immunol.* 58:102–118.

Hiwat, H., and G. Bretas. 2011. Ecology of *Anopheles darlingi* Root with respect to vector importance: a review. *Parasit. Vectors* 4:177.

Hlaing, T., et al. 2009. Mitochondrial pseudogenes in the nuclear genome of *Aedes aegypti* mosquitoes: implications for past and future population genetic studies. *BMC Genet.* 10:11.
Ho, B. C., A. Ewert, and L.-M. Chew. 1989. Interspecific competition among *Aedes aegypti*, *Ae. albopictus*, and *Ae. triseriatus* (Diptera: Culicidae): larval development in mixed cultures. *J. Med. Entomol.* 26:615–623.
Hochstrasser, M. 2023. Molecular biology of cytoplasmic incompatibility caused by *Wolbachia* endosymbionts. *Annu. Rev. Microbiol.* 2023:299–316.
Holt, R. A., et al. 2002. The genome of the malaria mosquito *Anopheles gambiae*. *Science* 298:129–149.
Hongoh, V., et al. 2012. Expanding geographical distribution of the mosquito, *Culex pipiens*, in Canada under climate change. *Appl. Geog.* 33:53–62.
Hopperstad, K. A., M. F. Sallam, and M. H. Reiskind. 2021. Estimation of fine-scale species distributions of *Aedes aegypti* and *Aedes albopictus* (Diptera: Culicidae) in eastern Florida. *J. Med. Entomol.* 58:699–707.
Hu, S. M. K. 1939. Preliminary observations on the effects of filarial infection on *Culex pipiens* var. *pallens* Coq. *Chinese Med. J.* 55:154–161.
Huang, J., R. Miller, and E. D. Walker. 2018. Cannibalism of egg and neonate larvae by late stage conspecific *Anopheles gambiae* (Diptera: Culicidae): implications for ovipositional studies. *J. Med. Entomol.* 55:801–807.
Huang, J., et al. 2007. The influence of darkness and visual contrast on oviposition by *Anopheles gambiae* in moist and dry substrates. *Physiol. Entomol.* 32:34–40.
Hubbard, A., et al. 2023. Implementing landscape genetics in molecular epidemiology to determine drivers of vector-borne disease: a malaria case study. *Mol. Ecol.* 32:1848–1859.
Huestis, D. L., et al. 2019. Windborne long-distance migration of malaria mosquitoes in the Sahel. *Nature* 10:404–408.
Hurd, H. 2003. Manipulation of medically important insect vectors by their parasites. *Ann. Rev. Entomol.* 48:141–161.
Isaacson, M. 1989. Airport malaria: a review. *Bull. World Health Organ.* 67:737–743.
Iwamura, T., A. Guzman-Holst, and K. A. Murray. 2020. Acclerating invasion potential of disease vector *Aedes aegypti* under climate change. *Nat. Commun.* 11:2130.
Iyaloo, D. P., et al. 2014. Guidelines to site selection for popualtion surveillance and mosquito control trials: a case study from Mauritius. *Acta Trop.* 132S:S140–S149.
Jackson, B. T., and S. L. Paulson 2006. Seasonal abundance of *Culex restuans* and *Culex pipiens* in southwestern Virginia through ovitrapping. *J. Am. Mosq. Contrl. Assoc.* 22:206–212.
Jackson, B. T., et al. 2005. Oviposition preferences of *Culex pipiens* and *Culex restans* (Diptera: Culicidae) for selected infusions in oviposition traps and gravid traps. *J. Am. Mosq. Contr. Assoc.* 21:360–365.

Jasper, M., et al. 2019. A genomic approach to inferring kinship reveals limited intergenerational dispersal in the yellow fever mosquito. *Mol. Ecol. Resour.* 92:1254–1265.

Jeannin, C., et al. 2022. An alien in Marseille: investigations on single *Aedes aegypti* mosquito likely introduced by a merchant ship from tropical Africa to Europe. *Parasite* 29:42.

Jobling, B. 1987. The mosquito *Aedes aegypti*. In *Anatomical Drawings of Biting Flies*. Prepared for publication by D. J. Lewis. London: British Museum (Natural History), in asociation with the Wellcome Trust.

Johnson, B. J., et al. 2020. Mosquito age grading and vector-control programmes. *Trends Parasitol.* 36:39–51.

Jones, C. E., et al. 2012. Rainfall influences survival of *Culex pipiens* (Dipetera: Culicidae) in a residential neighborhood in the mid-Atlantic United States. *J. Med. Entomol.* 49:467–473.

Joshi, V., D. T. Mourya, and R. C. Sharma. 2002. Persistence of dengue-3 virus through transovarial transmission passage in successive generations of *Aedes aegypti* mosquitoes. *Am. J. Trop. Med. Hyg.* 67:158–161.

Juarez, J. G., et al. 2020. Dispersal of female and male *Aedes aegypti* from discarded container habitats using a stable isotope mark-capture study design in South Texas. *Sci. Rep.* 10:6803.

Juliano, S. A. 1998. Species introduction and replacement among mosquitoes: interspecific resource competition or apparent competition? *Ecology* 79:255–268.

Juliano, S. A. 2000. Species interactions among larval mosquitoes: context dependence across habitat gradients. *Ann. Rev. Entomol.* 54:37–56.

Juliano, S. A., and L. Reminger. 1992. The relationship between vulnerability to predation and behavior of larval treehole mosquitoes: geographic and ontogenetic differences. *Oikos* 63:465–476.

Juliano, S. A., L. P. Lounibos, and G. F. O'Meara. 2004. A field test for competitive effects of *Aedes albopictus* on *A. aegypti* in South Florida: differences between sites of coexistence and exclsion? *Oecologia* 139:583–593.

Juneja, P., et al. 2014. Assembly of the genome of the disease vector *Aedes aegypti* onto a linkage map allows mapping of genes affecting disease transmission. *PLOS Negl. Trop. Dis.* 9:e2652.

Jupp, P. G., A. Kemp, and C. Frangos. 1991. The potential for dengue in South Africa: morphology and taxonomic status of *Aedes aegypti* populations. *Mosq. Syst.* 23:182–190.

Kamau, L., et al. 1998. Microgeographic genetic differentiation of *Anopheles gambiae* from Asembo Bay, Western Kenya: a comparison with Kelifi in costal Kenya. *Am. J. Trop. Med. Hyg.* 58:64–69.

Kaufman, M. G., et al. 2006. Importance of algal biomass to growth and development of *Anopheles gambiae* larvae. *J. Med. Entomol.* 43:669–676.

Kefi, M., et al. 2024. Curing mosquitoes with genetic approaches for malaria control. *Trends Parasitol.* 40:487–499.

Kenea, O., et al. 2017. Comparison of two adult mosquito sampling methods with human landing catches in south-central Ethiopia. *Malar. J.* 16:30.

Kennedy, D., and M. Lucks. 1999. Rubber, blight, and mosquitoes: biogeography meets the global economy. *Environ. Hist.* 4:369–383.

Kent, T. V., J. Uzunovic, and S. I Wright 2017. Coevolution between transposable elements and recombination. *Roy. Soc. Phil. Trans. B* 372:20160458.

Keyghobadi, N., et al. 2006. Fine-scale population genetic structure of a wildlife disease vector: the southern house mosquito on the island of Hawaii. *Mol. Ecol.* 10:1111.

Khalil, N., et al. 2021. Host associations of *Culex pipiens*: a two-year analysis of bloodmeal sources in sotheastern Virginia. *Vector-Borne Zoo. Dis.* 21:961–972.

Khatchikian, C. E., et al. 2009. Climate and geographic trends in hatch delay of the treehole mosquito, *Aedes triseriatus* Say (Diptera: Culicidae). *J. Vector Ecol.* 34:119–128.

Khin, M. M., and K. A. Than. 1983. Transovarial transmission of dengue-2 virus by *Aedes aegypti* in nature. *Am. J. Trop. Med. Hyg.* 32:590–594.

Kim, S., S. Trocke, and C. Sim. 2018. Comparative studies of stenogamous behavior in the mosquito *Culex pipiens* complex. *Med. Vet. Entomol.* 32:427–435.

Kirkpatrick, M. 2010. How and why chromosome inversions evolve. *PLOS Biol.* 8:e1000501.

Kitzmiller, J. B., and G. F. Mason. 1967. Formal genetics of anophelines. Pp. 3–16 in J. W. Wright and R. Pal, eds., *Genetics of Insect Vectors of Disease*. Amsterdam: Elsevier.

Klaus, A. V., V. L. Kulasekera, and V. Schawaroch. 2002. Three-dimensional visualization of insect morphology using confocal laser scanning microscopy. *J. Microsc.* 212:107–121.

Kloke, R. G. 1997. New distribution record of *Anopheles merus* Donitz (Diptera: Culicidae) in Zambia. *African Entomol.* 5:361–362.

Knab, F. 1906. The swarming of *Culex pipiens*. *Psyche* 13:123–133.

Koenekoop, R., and T. Livdahl. 1986. Cannibalism among *Aedes triseriatus* larvae. *Ecol. Entomol.* 11:111–114.

Koenraadt, C. J. M., and W. Takken. 2003. Cannibalism and predation among larvae of the *Anopheles gambiae* complex. *Med. Vet. Entomol.* 17:61–67.

Koenraadt, C. J. M., et al. 2004. The effects of food and space on the occurrence of cannibalism and predation among larvae of *Anopheles gambiae* s.l. *Entomol. Exp. Appl.* 112:125–134.

Kothera, L., et al. 2012. Complexity of the *Culex pipiens* complex in California. *Proc. Pap. Annu. Conf. Mosq. Vector Control Assoc. Calif.* 80:1–3.

Kothera, L., et al. 2013. Population genetic and admixture analyses of *Culex pipiens* complex (Diptera: Culicidae) populations in California, USA. *Am. J. Trop. Med. Hyg.* 89:1154–1167.

Kothera, L., et al. 2020. Blood meal, host selection, and genetic admixture analyses of *Culex pipiens* complex (Diptera: Culicidae) mosquitoes in Chicago, IL. *J. Med. Entomol.* 57:78–87.

Kotsakiozi, P., et al. 2017a. Population genomics of the Asian tiger mosquito, *Aedes albopictus*: insights into the recent worldwide invasion. *Ecol. Evol.* 7:10143–10157.

Kotsakiozi, P., et al. 2017b. Tracking the return of *Aedes aegypti* to Brazil, the major vector of the dengue, chikungunya and Zika viruses. *PLOS Negl. Trop. Dis.* 11:e0005663.

Kotsakiozi, P., et al. 2018a. *Aedes aegypti* in the Black Sea: recent introduction or ancient remnant? *Parasit. Vectors* 11:396.

Kotsakiozi, P., et al. 2018b. Population structure of a vector of human diseases: *Aedes aegypti* in its ancestral range, Africa. *Ecol. Evol.* 8:7835–7848.

Kraemer, M. U. G., et al. 2019. Past and future spread of the arbovirus vectors *Aedes aegypti* and *Aedes albopictus. Nat. Microbiol.* 4:854–863.

Krimbas, C. B., and J. R. Powell, eds. 1992. Drosophila *Inversion Polymorphism.* Boca Raton, FL: CRC Press.

Krockel, U. A., et al. 2006. New tools for surveillance of adult yellow fever mosquitoes: comparison of trap catches with human landing rates in an urban environment. *J. Am. Mosq. Contr. Assoc.* 22:229–238.

Kuno, G. 2010. Early history of laboratory breeding of *Aedes aegypti* (Diptera: Culicidae) focusing on the origin of selected strains. *J. Med. Entomol.* 47:957–971.

Lahondere, C., et al. 2020. The olfactory basis of orchid pollination by mosquitoes. *Proc. Natl. Acad. Sci. USA* 117:708–716.

Lalonde, M. M., and J. M. Marcus. 2020. How old can we go?: evaluating the age limit for effective DNA recovery from historical insect specimens. *Syst. Entomol.* 45:505–515.

Lanciotti, R. S., et al. 1999 Origin of the West Nile virus responsible for an outbreak of encephalitis in the northeastern United States. *Science* 286:2333–2337.

Lang, C. A., H. Y Lau, and D. J. Jefferson. 1965. Protein and nucleic acid changes during growth and aging in mosquitoes. *Biochem. J.* 95:372–377.

Lanzaro, G. C., et al. 1995. Microsatellite DNA and isozyme variability in a West African population of *Anopheles gambiae. Insect Mol. Biol.* 4:105–112.

Lanzaro, G. C., et al. 1998. Complexities in the genetic structure of *Anopheles gambiae* populations in West Africa as revealed by microsatellite analysis. *Proc. Natl. Acad, Sci. USA* 95:14260–14265.

Lapointe, D. A. 2008. Dispersal of *Culex quinquefasciatus* (Diptera: Culicidae) in a Hawaiian rain forest. *J. Med. Entomol.* 45:600–609.

Laporta, G. Z., and M. A. M. Sallum. 2008. Density and survival rate of *Culex quinquefasciatus* in Parque Ecológico do Tietê, São Paulo, Brazil. *J. Am. Mosq. Contr. Assoc.* 24:21–27.

Laporta, G. Z., et al. 2023. Global distribution of *Aedes aegypti* and *Aedes albopictus* in a climate change scenario of regional rivalry. *Insects* 14:010049.

Laurence, B. R., and J. A. Pickett. 1985. An oviposition attractant pheromone in *Culex quinquefasciatus* Say (Diptera: Culicidae). *Bull. Ent. Res.* 75:283–290.

Laven, H. 1967. Formal genetics of *Culex pipiens*. Pp. 17–65 in J. W. Wright and R. Pal, eds., *Genetics of Insect Vectors of Disease*. Amsterdam: Elsevier.

Lawniczak, M. K. N., et al. 2010. Incipient *Anopheles gambiae* species revealed by whole genome sequences. *Science* 330:512–514.

Lawson, D. J., L. van Dorp, and D. Falush. 2018. A tutorial on how not to over-interpret STRUCTURE and ADMIXTURE bar plots. *Nat. Comm.* 9:3258.

Leal, W. S., et al. 2008. Reverse and conventional chemical ecology approaches for the development of oviposition attractants for *Culex* mosquitoes. *PLOS ONE* 3:e3045.

Lee, H. L., and A. Rohani. 2005. Transovarial transmission of dengue virus in *Aedes aegypti* and *Aedes albopictus* in relation to dengue outbreak in an urban area in Malaysia. *Dengue Bull.* 29:106–111.

Lefèvre, T., et al. 2009a. Beyond nature and nurture: Phenotype plasticity in blood-feeding behavior of *Anopheles gambiae* s.s. when humans are not readily accessible. *Am. J. Trop. Med. Hyg.* 81:1023–1029.

Lefèvre, T., et al. 2009b. Evolutionary lability of odour-mediated host preference by the malaria vector *Anopheles gambiae*. *Trop. Med. Internat. Health* 14:228–236.

Lefèvre, T., et al. 2010. Beer consumption increases human attractiveness to malaria mosquitoes. *PLOS One* 5:e9546.

Le Goff, G., C. Brengues, and V. Robert. 2013. *Stegomyia* mosquitoes in Mayotte, taxonomic study and description of *Stegomyia pia* n. sp. *Parasite* 20:31.

Lehane, M. J. 2005. *The Biology of Blood-Sucking Insects*. Cambridge: Cambridge University Press.

Lehmann, T., and A. Diabate 2008. Molecular forms of *Anopheles gambiae*: a phenotypic perspective. *Infec. Genet. Evol.* 8:737–746.

Lehmann, T., et al. 1996. Genetic differentiation of *Anopheles gambiae* populations from East and West Africa: comparison of microsatellite and allozyme loci. *Heredity* 77:192–200.

Lehmann, T., et al. 1997. Microgeographic structure of *Anopheles gambiae* in western Kenya based on mtDNA and microsatellite loci. *Mol. Ecol.* 6:243–253.

Lehmann, T., et al. 1999. The Rift Valley complex as a barrier to gene flow for *Anopheles gambiae* in Kenya. *J. Hered.* 90:613–621.
Lehmann, T., et al. 2003. Population structure of *Anopheles gambiae* in Africa. *J. Hered.* 94:133–147.
Lehmann, T., et al. 2010. Aestivation of the African malaria mosquito *Anopheles gambiae* in the Sahel. *Am. J. Trop. Med. Hyg.* 83:601–606.
Lehmann, T., et al. 2014. Seasonal variation in spatial distributions of *Anopheles gambiae* in a Sahelian village: evidence for aestivation. *J. Med. Entomol.* 51:27–38.
Leisnham, P. T., et al. 2021. Condition-specific competitive effects of the invasive mosquito *Aedes albopictus* on the resident *Culex pipiens* among different urban container habitats may explain their coexistence in the field. *Insects* 12:993.
Lewontin, R. C. 1974. *The Genetic Basis of Evolutionary Change*. New York: Columbia University Press.
Li, C., et al. 2021. Potential geographic distributon of *Anopheles gambiae* worldwide under climate change. *J. Biosafety Biosec.* 3:125–130.
Li, J., et al. 2009. Effects of water color and chemical compounds on the oviposition behavior of gravid *Culex pipiens pallens* females under laboratory conditions. *J. Agric. Urban Entomol.* 26:23–30.
Li, Y., et al. 2016. Comparative evaluation of the efficiency of the BG-sentinel trap, CDC light trap and mosquito-ovioposition trap for the surveillance of vector mosquitoes. *Parasit. Vectors* 9:446.
Liang, J., et al. 2024. Discovery and characterization of chromosomal inversions in the arboviral vector mosquito *Aedes aegypti*. *bioRχiv* 2.16.580682. https://doi.org/10.1101/2024.02.16.580682.
Lima, C. A., et al. 2003. Reproductive aspects of the mosquito *Culex quinquefasciatus* (Diptera: Culicidae) infected with *Wuchereria bancrofti* (Spirurida: Onchocercidae). *Mem. Inst. Oswald Cruz* 98:217–222.
Lindh, J. M., et al. 2008. Oviposition response of *Anopheles gambiae* s.s. (Diptera: Culicidae) and identification of volatiles from bacteria-containing solutions. *J. Med. Entomol.* 45:1039–1049.
Lindh, J. M., et al. 2015. Discovery of an oviposition attractant for gravid malaria vectors of the *Anopheles gambiae* species complex. *Malar. J.* 14:119.
Lindquist, A. W., et al. 1967. Dispersion studies of *Culex pipiens fatigans* tagged with ^{32}P in the Kemmendine area of Rangoon, Burma. *Bull. World Health Organ.* 36:21–37.
Linthicum, K. J., et al. 2003. Introduction and potential establishment of *Aedes albopictus* in California in 2001. *J. Am. Mosq. Contr. Assoc.* 19:301–308.
Liu, W., et al. 2023. Chromosome-level assembly of *Culex pipiens molestus* and improved reference genome of *Culex pipiens pallens* (Culicidae, Diptera). *Mol. Ecol. Resour.* 23:486–498.

Liu-Helmersson, A., et al. 2018. Estimating past, present, and future trends in the global distribution and abundance of the arbovirus vector *Aedes aegypti* under climate change scenarios. *Front. Public Health* 7:148.

Livdahl, T. P., and M. S. Wiley 1991. Prospects for an invasion: competition between *Aedes albopictus* and native *Aedes triseriatus*. *Science* 251:189–191.

Lounibos, L. P. 1981. Habitat segregation among African treehole mosquitoes. *Ecol. Entomol.* 6:129–154.

Lounibos, L. P. 2002. Invasion of insect vectors of human disease. *Ann. Rev. Entomol.* 47:233–266.

Lounibos, L. P. 2003. Genetic-control trials and the ecology of *Aedes aegypti* at the Kenya coast. Pp. 33–46 in W. Takken and T. W. Scott, eds., *Ecological Aspects for Application of Genetically Modified Mosquitoes*. Dordrecht, Netherlands: Kluwer Academic.

Lounibos, L. P., and S. A. Juliano. 2018. Where vectors collide: the importance of mechanisms shaping the realized niche for modeling ranges of invasive *Aedes* mosquitoes. *Biol. Invasions* 20:1913–1929.

Lounibos, L. P., J. R. Rey, and J. H. Frank. 1985. *Ecology of Mosquitoes: Proceedings of a Workshop*. Vero Beach: Florida Medical Entomology Laboratory.

Lovin, D. D., et al. 2009. Genome-based polymorphic microsatellite development and validation in the mosquito *Aedes aegypti* and application to population genetics in Haiti. *BMC Genet.* 10:590.

Lumsden, W. H. R. 1957. The activity cycle of domestic *Aedes* (*Stegomyia*) *aegypti* (L.) (Dipt., Culicid.) in Southern Province, Tanganyika. *Bull. Ent. Res.* 48:769–782.

Lyimo, E. O., and W. Takken. 1993. Effects of adult body size on fecundity and the pre-gravid rate of *Anopheles gambiae* in Tanzania. *Med. Vet. Entomol.* 7:328–332.

MacDonald, W. W., A. Sebastian, and M. M. Tun. 1968. A mark-recapture experiment with *Culex pipiens fatigans*. *Ann. Trop. Med. Parasitol.* 62:200–209.

MacGregor, M. E. 1924. *Aedes* (*Stegomyia*) *mascarensis* MacGregor: a new mosquito from Mauritius. *Bull. Entomol. Res.* 14:409–412.

Maciel-de-Freitas, C., T. Codeço, and R. Lourenço-de-Oliveira. 2007a. Body size-associated survival and dispersal rates of *Aedes aegypti* in Rio de Janeiro. *Med. Vet. Entomol.* 21:284–292.

Maciel-de-Freitas, C., T. Codeço, and R. Lourenço-de-Oliveira. 2007b. Daily survival rates and dispersal of *Aedes aegypti* females in Rio de Janiero, Brazil. *Am. J. Trop. Med. Hyg.* 76:659–665.

Madan, D., et al. 2022. Estimating female malaria mosquito age by quantifying Y-linked genes in stored male spermatozoa. *Sci. Rep.* 12:10570.

Mahon, R. J., C. A. Green, and R. H. Hunt. 1976. Diagnostic allozymes for routine identification of adults of the *Anopheles gambiae* complex (Diptera, Culicidae). *Bull. Entomol. Res.* 66:25–31.

Maitra, A., et al. 2019. Exploring deeper genetic structures: *Aedes aegypti* in Brazil. *Acta Trop.* 195:68–77.

Mala, A. O., and L. W. Irungu. 2011. Factors influencing differential larval habitat productivity of *Anopheles gambiae* complex in a western Kenyan village. *J. Vector Borne Dis.* 48:52–57.

Mallet, J., N. J. Besansky, and M. W. Hahn. 2016. How reticulated are species? *Bioassays* 38:140–149.

Mancini, M. V., and G. Favia. 2022. *Asaia* pratransgenesis in mosquitoes. Pp. 308–327 in M. Q. Benedict and M. J. Scott, eds., *Transgenic Insects*. Boston: CAB International.

Mancini, M. V., et al. 2018. Estimating bacteria diversity in different organs of nine species of mosquito by next generation sequencing. *BMC Microbiol.* 18:126.

Manni, M., et al. 2017. Genetic evidence for a worldwide chaotic dispersion pattern of the arbovirus vector, *Aedes albopictus. PLOS Negl. Trop. Dis.* 11:0005332.

Manoukis, N., et al. 2009. Structure and dynamics of male swarms of *Anopheles gambiae. J. Med. Entomol.* 46:227–235.

Marcantonio, M., T. Reyes, and C. M. Barker. 2019. Quantifying *Aedes aegypti* dispersal in space and time: a modeling approach. *Ecosphere* 10:e02977.

Marini, G., et al. 2016. The role of climatic and density dependent factors in shaping mosquito population dynamics: the case of *Culex pipiens* in northwestern Italy. *PLOS ONE* 11:e154018.

Marini, G., et al. 2017. The effect of interspecific competion on the temporal dynamics of *Aedes albopictus* and *Culex pipiens. Parsit. Vectors* 10:102.

Marinotti, O., et al. 2013. The genome of *Anopheles darlingi*, the main neotropical malaria vector. *Nucl. Acids Res.* 41:7387–7400.

Marten, G. G., and J. W. Reid. 2007. Cyclopoid copepods. *J. Am. Mosq. Contr. Assoc.* 23:65–92.

Matthews, B. J., M. A. Younger, and L. B. Vosshall. 2019. The ion channel *ppk301* controls freshwater egg-laying in the mosquito *Aedes aegypti. eLife* 8:43963.

Matthews, B. J., et al. 2018. Improved *Aedes aegypti* mosquito reference genome assembly enables biological discovery and vector control. *Nature* 563:501–507.

Matthews, J., A. Bethel, and G. Osei. 2020. An overview of malarial *Anopheles* mosquito survival estimates in relation methodology. *Parast. Vectors* 13:233.

Mattingly, P. F. 1957. Genetical aspects of the *Aedes aegypti* problem, I: taxonomy and bionomics. *Ann. Trop. Med. Parasitol.* 51:392–408.

Mbaika, S., et al. 2016. Vector competence of *Aedes aegypti* in transmitting chikungunya virus: effects and implications of extrinsic incubation temperature on dissemination and infection rates. *Virology J.* 13:114.

Mburu, M. M., et al. 2019. Assessment of the Suna trap for sampling mosquitoes indoors and outdoors. *Malar. J.* 18:51.

McAbee, R. D., J. A. Christiansen, and A. J. Cornel. 2007. Detailed larval salivary gland polytene chromosome photomap for *Culex quinquefasciatus* (Diptera: Culicidae) from Johannesburg, South Africa. *J. Med. Entomol.* 44:229–237.

McBride, C. S. 2016. Genes and odors underlying the recent evolution of mosquito preference for humans. *Curr. Biol.* 26:R41–R46.

McBride, C. S., et al. 2014. Evolution of mosquito preference for humans linked to an odorant receptor. *Nature* 515:222–227.

McClelland, G. A. H. 1974. A worldwide survey of variation in scale pattern of the abdominal tergum of *Aedes aegypti* (L.) (Diptera: Culicidae). *Trans. R. Entomol. Soc. Lond.* 126:239–259.

McCrae, A. W. 1984. Oviposition by African malaria vector mosquitoes, II: effects of site tone, water type and conspecific immatures on target selection by freshwater *Anopheles gambiae* Giles, sensu lato. *Ann. Trop. Med. Parasitol.* 78:307–318.

McDonald, P. T. 1977. Population characteristics of domestic *Aedes aegypti* (Diptera: Culicidae) in villages on the Kenya coast, I: adult survivorship and population size. *J. Med. Entomol.* 14:42–48.

McIver, S. B., T. J. Wilkes, and M. T. Gillies 1980. Attraction to mammals of male *Mansonia* (*Mansonioides*) (Diptera: Culicidae). *Bull. Entomol. Res.* 70:11–16.

McMeniman, C. J., et al. 2009. Stable introduction of a life-shortening *Wolbachia* infection into the mosquito *Aedes aegypti*. *Science* 323:141–144.

McNeill, J. R. 2010. *Mosquito Empires*. Cambridge: Cambridge University Press.

Merritt, R. W., R. H. Dadd, and E. D. Walker. 1992. Feeding behaviour, natural food and nutritional relationships of larval mosquitoes. *Annu. Rev. Entomol.* 37:349–376.

Meuti, M. E., C. A. Short, and D. L. Denlinger. 2015. Mom matters: diapause characteristics of *Culex pipiens-Culex quinquefasciatus* (Diptera: Culicidae) hybrid mosquitoes. *J. Med. Entomol.* 52:131–137.

Midega, J. T., et al. 2007. Estimating dispersal and survival of *Anopheles gambiae* and *Anopheles funestus* along the Kenyan coast by using mark-release-recapture methods. *J. Med. Entomol.* 44:923–929.

Mihou, A. P., and A. Michaelakis. 2010. Oviposition aggregation pheromone for *Culex* mosquitoes: bioactivity and synthetic approaches. *Hell. Plant Prot. J.* 3:33–56.

Miller, J. R., et al. 2007. Life on the edge: African malaria mosquitoes (*Anopheles gambiae* s.l.) are amphibious. *Naturwissen.* 94:195–199.

Mitchell, C. J., and H. Briegel. 1989. Inability of diapausing *Culex pipiens* (Diptera: Culicidae) to use blood for producing lipid reserves for overwinter survival. *J. Med. Entomol.* 26:318–326.

Moll, R. M., et al. 2001. Meconial peritrophic membranes and the fate of midgut bacteria during mosquito (Diptera: Culicidae) metamorphosis. *J. Med. Entomol.* 38:29–32.

Monteiro, F. A., et al. 2014. Genetic diversity of Brazilian *Aedes aegypti*: patterns following an eradication program. *PLOS Negl. Trop. Dis.* 8:e3167.

Moore, C. G. 1999. *Aedes albopictus* in the United States: current status and prospects for further spread. *J. Am. Mosq. Contr. Assoc.* 15:221–227.

Moore, D. 1979. Hybridization and mating behavior of *Aedes aegypti* (Diptera: Culicidae). *J. Med. Entomol.* 16:223–226.

Mordecai, W. A., et al. 2020. Climate change could shift disease burden from malaria to arboviruses in Africa. *Lancet Planet Health* 4:e416–e423.

Moreira, L. A., et al. 2009. A *Wolbachia* symbiont in *Aedes aegypti* limits infection with dengue, chikungunya, and *Plasmodium*. *Cell* 139:1268–1278.

Morinaga, G., et al. Submitted. Comparative genomics based on *de novo* genome assemblies of *Aedes aegypti formosus* and *Aedes mascarensis* from wild specimens.

Mori, A., J. Romero-Severson, and D. W. Severson. 2007. Genetic basis for reproductive diapause is correlated with life history traits within the *Culex pipiens* complex. *Insect Mol. Biol.* 16:515–524.

Mori, A., D. W. Severson, and B. M. Christensen. 1999. Comparative linkage maps for the mosquitoes (*Culex pipiens* and *Aedes aegypti*) based on common RFLP loci. *J. Hered.* 90:161–164.

Morrison, A. C., et al. 2008. Defining challenges and proposing solutions for control of the virus vector *Aedes aegypti*. *PLOS Med.* 5:e68.

Mousson, L., et al. 2001. Genetic structure of *Aedes aegypti* populations in Chiang Mai (Thailand) and relation with dengue transmission. *Trop. Med. Int. Health* 7:865–872.

Motoki, M. T., et al. 2019. Population genetics of *Aedes albopictus* (Diptera: Culicidae) in its native range in Lao People's Democratic Republic. 2019. *Parasit. Vectors* 12:477.

Muir, L. E., and B. H. Kay. 1998. *Aedes aegypti* survival and dispersal estimated by mark-release-recapture in northern Australia. *Am. J. Trop. Med. Hyg.* 58:277–283.

Munga, S., et al. 2005. Oviposition site preference and egg hatchability of *Anopheles gambiae*: effects of land cover types. *J. Med. Entomol.* 42:993–997.

Munga, S., et al. 2006. Effects of larval competitors and predators on oviposition site selection of *Anopheles gambiae* sensu stricto. *J. Med. Entomol.* 43:221–224.

Munstermann, L. 1994. Gene map of the yellow fever mosquito *Aedes* (*Stegomyia*) *aegypti* (2N=6). Pp. 3264–3268 in S. J. O'Brien, ed., *Genetic Maps: Locus Maps of Complex Genomes*, 6th edition. Cold Spring Harbor, NY: Cold Spring Harbor Laboratory.

Munstermann, L., and G. B. Craig Jr. 1979. Genetics of *Aedes aegypti*: updating the linkage map. *J. Hered.* 70:291–294.

Mustafa, M. S., et al. 2021. Population genetics of *Anopheles arabiensis* the primary malaria vector in the Republic of Sudan. *Malar. J.* 20:469.

Muturi, E. J., et al. 2010. Population genetic structure of *Anopheles arabiensis* (Diptera: Culicidae) in a rice growing area of central Kenya. *J. Med. Entomol.* 47:144–151.

Mwanga, E. P., et al. 2019. Evaluation of the ultraviolet LED trap for catching *Anopheles* and *Culex* mosquitoes in south-eastern Tanzania. *Parasit. Vectors* 12:418.

Nathan, M. B., D. A. Focks, and A. Kroeger. 2006. Pupal/demographic surveys to inform dengue-vector control. *Ann. Trop. Med. Parasit.* 100:S1–S3.

Neafsey, D. E., et al. 2010. SNP genotyping defines complex gene-flow boundaries among African malaria vector mosquitoes. *Science* 330:514–517.

Ng'habi, K. R., et al. 2011. Population genetic structure of *Anopheles arabiensis* and *Anopheles gambiae* in a malaria endemic region of southern Tanzania. *Malar. J.* 10:289.

Nguyen, P. L., et al. 2017. No evidence for manipulation of *Anopholes gambiae*, *An. coluzzii* and *An. arabiensis* host preference by *Plasmodium falciparum*. *Sci. Rep.* 7:9415.

Niang, A., et al. 2022. Perfect association between spatial swarm segregation and the X-chromosome speciation island in hybridizing *Anopheles coluzzii* and *Anopheles gambiae* populations. *Sci. Rep.* 12:10800.

Noori, N., B. G. Lockaby, and L. Kalin. 2015. Larval development of *Culex quinquefasciatus* in water with low to moderate pollution levels. *J. Vector Ecol.* 40:208–220.

Norris, L. C., et al. 2010. Frequency of multiple blood meals taken in a single gonotrophic cycle by *Anopheles gambiae* in Macha, Zambia. *Am. J. Trop. Med. Hyg.* 83:33–37.

Nwakanma, D. C., et al. 2013. Breakdown in the process of incipient speciation in *Anopheles gambiae*. *Genetics* 193:1221–1231.

Nyanjom, S. R., et al. 2003. Population genetic structure of *Anopheles arabiensis* mosquitoes in Ethiopia and Eritrea. *J. Hered.* 94:457–463.

Nyasembe, V. O., and B. Torto. 2014. Volatile phytochemicals as mosquito semiochemicals. *Phytochem. Lett.* 8:196–201.

Odero, J. O., et al. 2019. Using sibship reconsructions to understand the relationship between larval habitat productivity and oviposition behaviour in Kenyan *Anopheles arabiensis*. *Malar. J.* 18:286.

Ogbunugafor, C. B., and L. Suma. 2008. Behavioral evidence for the existence of region-specific oviposition cue in *Anopheles gambiae* s.s. *J. Vector Ecol.* 33:321–324.

Olanratmanee, P., et al. 2013. Population genetic structure of *Aedes* (*Stegomyia*) *aegypti* (L.) at a microspatial scale in Thailand: implications for dengue suppression. *PLOS Negl. Trop. Dis.* 7:31913.

Omer, S. M., and J. L. Cloudsley-Thompson. 1970. Survival of female *Anopheles gambiae* Giles through a 9-month dry season in Sudan. *Bull. World Health Organ.* 42:319–330.

Onyabe, D. Y., and J. E. Conn. 2001a. Genetic differentiation of the malaria vector *Anopheles gambiae* across Nigeria suggests that selection limits gene flow. *Heredity* 87:647–658.

Onyabe, D. Y., and J. E. Conn. 2001b. Population genetic structure of the malaria mosquito *Anopheles arabiensis* across Nigeria suggests range expansion. *Mol. Ecol.* 10:2577–2591.

Onyeka, J. O. A., and P. F. L. Boreham. 2009. Population studies, physiological state and mortality factors of overwintering adult populations of females of *Culex pipiens* L. (Diptera: Culicidae). *Bull. Entomol. Res.* 77:99–111.

Orsborne, J., et al. 2018. Using the human blood index to investigate host biting plasticity: a systematic review and meta-regression of the three major African malaria vectors. *Malar. J.* 17:479.

Osgood, C. E. 1971. An oviposition pheromone associated with egg rafts of *Culex tarsalis*. *J. Econ. Entomol.* 64:1038–1041.

Otero, M., H. G. Solari, and N. Schweigmann. 2006. A stochastic population dynamics model for *Aedes aegypti*: formulation and application to a city with temperate climate. *Bull. Math. Biol.* 68:1945–1974.

Ouedraogo, W. M., et al. 2022. Impact of physicochemical parameters of *Aedes aegypti* breeding habitats on mosquito productivity and the size of emerged adult mosquitoes in Ouagadougou City, Burkina Faso. *Parasit. Vectors* 15:478.

Palatini, U., et al. 2020. Improved reference genome of the arboviral vector *Aedes albopictus*. *Genome Biol.* 21:215.

Pan, X., et al. 2018. The bacterium *Wolbachia* exploits host innate immunity to establish a symbiotic relationship with the dengue vector mosquito *Aedes aegypti*. *ISME J.* 12:277–288.

Pardue, M. L., and J. G. Gall. 1970. Chromosomal localization of mouse satellite DNA. *Science* 168:1356–1358.

Parmakelis, A., et al. 2008. Historical analysis of a near disaster: *Anopheles gambiae* in Brazil. *Am. J. Trop. Med. Hyg.* 78:176–178.

Patterson, H. E. 1964. "Saltwater *Anopheles gambiae*" on Mauritius. *Bull. World Health Organ.* 31:635–644.

Paupy, C., et al. 2000. *Aedes aegypti* in Tahiti and Moorea (French Polynesia): isoenzyme differentiation in the mosquito population according to human population density. *Am. J. Trop. Med. Hyg.* 62:217–224.

Paupy, C., et al. 2012. Genetic structure and phylogeography of *Aedes aegypti*, the dengue and yellow-fever mosquito vector in Bolivia. *Infec. Genet. Evol.* 12:1260–1269.

Peach, D. A. H., and G. Gries. 2020. Mosquito phytophagy—sources exploited, ecological function, and evolutionary transition to haematophagy. *Entomol. Exp. App.* 168:120–136.

Peirce, M. J., et al. 2020. JNK signaling regulates oviposition in the malaria vector *Anopheles gambiae*. *Sci. Rep.* 10:14344.

Peris, D., et al. 2020. DNA from resin-embedded organisms: past, present and future. PLOS ONE 15:e0239521.

Peterson, A. T., and L. P. Campbell. 2015. Global potential distribution of the mosquito *Aedes notoscriptus*, a new alien species in the United States. *J. Vector Ecol.* 40:191–194.

Petrarca, V., and J. C. Beier. 1992. Intraspecific chromosomal polymorphism in the *Anopheles gambiae* complex as a factor affecting malaria transmission in the Kisumu area of Kenya. *Am. J. Trop. Med. Hyg.* 46:229–237.

Pichler, V., et al. 2019. Complex interplay of evolutionary forces shaping population genomic structure of invasive *Aedes albopictus* in southern Europe. *PLOS Negl. Trop. Dis.* 13:e0007554.

Pickrell, J. K., and J. K. Pritchard. 2012. Inference of population splits and mixtures from genome-wide allele frequency data. *PLOS Genet.* 8:e1002967.

Platt, K. B., et al. 1997. Impact of dengue virus infection on feeding behavior of *Aedes aegypti*. *Am. J. Trop. Med. Hyg.* 57:119–125.

Pless, E., and V. Raman. 2018. Origin of *Aedes aegypti* in Clark County, Nevada. *J. Am. Mosq. Contr. Assoc.* 34:302–305.

Pless, E., et al. 2017. Multiple introductions of the dengue vector, *Aedes aegypti*, into California. *PLOS Negl. Trop. Dis.* 11:e0005718.

Pless, E., et al. 2020. Sunshine versus gold: the effect of population age on genetic structure of an invasive mosquito. *Ecol. Evol.* 10:9588–9599.

Pless, E., et al. 2021. A machine learning approach to map landscape connectivity in *Aedes aegypti* with genetic and environmental data. *Proc. Natl. Acad. Sci. USA* 118:e2003201118.

Pless, E., et al. 2022. Evidence for serial founder events during the colonization of North America by the yellow fever mosquito, *Aedes aegypti*. *Ecol. Evol.* 12:e8896.

Pohl, K., and I. A. Cockburn. 2022. Innate immunity to malaria: the good, the bad and the unknown. *Front. Immunol.* 13:914598.

Poinar, J. O., et al. 2000. *Paleoculicis minutus* (Diptera: Culicidae) n. gen., n. sp., from Cretaceous Canadian amber, with a summary of described fossil mosquitoes. *Acta Geologica Hispanica* 35:119–128.

Pombi, M. P., et al. 2008. Chromosomal plasticity and evolutionary potential in the malaria vector *Anopheles gambiae* sensu stricto: insights from three decades of rare paracentric inversions. *BMC Evol. Biol.* 8:309.

Pombi, M. P., et al. 2017. Dissecting functional components of reproductive isolation among closely related sympatric species of the *Anopheles gambiae* complex. *Evol. Appl.* 10:1102–1120.

Ponlawat, A., and L. C. Harrington. 2005. Blood feeding patterns of *Aedes aegypti* and *Aedes albopictus* in Thailand. *J. Med. Entomol.* 42:844–849.

Ponnusamy, L., et al. 2008. Identification of bacteria and bacteria-associated chemical cues that mediate oviposition site preferences by *Aedes aegypti*. *Proc. Natl. Acad. Sci. USA* 105:9262–9267.

Porretta, D., et al. 2016. Intra-instar larval cannibalism in *Anopheles gambiae* (s.s.) and *Anopheles stephensi* (Diptera: Culicidae). *Parasit. Vectors* 9:566–574.

Powell, J. R. 1994. Molecular techniques in population genetics: a brief history. Pp.131–156 in B. Shierwater et al., eds., *Molecular Approaches to Ecology and Evolution*. Basel: Birkhauser.

Powell, J. R. 2016. Mosquitoes on the move. *Science* 354:971–972.

Powell, J. R. 2018a. Genetic heterogeneity of insect vectors: death of typology? *Insects* 9:139.

Powell, J. R. 2018b. Mosquito-borne viral human diseases: why *Aedes aegypti*? *Am. J. Trop. Med. Hyg.* 98:1563–1565.

Powell, J. R. 2019. An evolutionary perspective on vector-borne diseases. *Front. Genet.* 10:1266.

Powell, J. R. 2022. Modifying mosquitoes to suppress disease transmission: is the long wait over? *Genetics* 221:iyac072.

Powell, J. R., and W. J. Tabachnick. 2014. Genetic shifting: a novel approach for controlling vector-borne diseases. *Trends Parasitol.* 30:282–288.

Powell, J. R., A. Gloria-Soria, and P. Kotsakiozi. 2018. Recent history of *Aedes aegypti*: vector genomics and epidemiology records. *BioScience* 68:854–860.

Powell, J. R., W. Tabachnick, and J. Arnold. 1980. Genetics and the origin of a vector population: *Aedes aegypti*, a case study. *Science* 208:1385–1387.

Powell, J. R., et al. 1999. Population structure, speciation, and introgression in the *Anopheles gambiae* complex. *Parassitologia* 41:101–113.

Powell, J. R., et al. 2014. Mario Coluzzi (1938–2012). *Malar. J.* 13:10.

Prince, B. C., et al. 2023. Recognition of arboviruses by the mosquito immune system. *Biomolecules* 13:1159.

Pritchard, J. K., M. Stephens, and P. Donnelly. 2000. Inference of population structure using mutilocus genotype data. *Genetics* 155:945–959.

Provine, W. B. 1971. *The Origins of Theoretical Population Genetics*. Chicago: University of Chicago Press.

Puchot, N., et al. 2022. Establishment of a colony of *Anopheles darlingi* from French Guiana for vector competence studies on malaria transmission. *Front. Trop. Dis.* 3:949300.

Quiroz-Martinez, H., and A. Rodriguez-Castro. 2007. Aquatic insects as predators of mosquito larvae. *J. Am. Mosq. Contr. Assoc.* 23 (Suppl. 2):110–117.

Ramasamy, R., et al. 2014. Biological differences between brackish and fresh water-derived *Aedes aegypti* from two locations in the Jaffna Peninsula of Sri Lanka and the implications for arboviral disease transmission. *PLOS ONE* 9: e104977.

Rances, E., et al. 2012. The relative importance of innate immune priming in *Wolbachia*-mediated dengue interference. *PLOS Pathog.* 8:e1002548.

Rasic, G., et al. 2015. *Aedes aegypti* has spatially structured and seasonally stable populations in Yogyakarta, Indonesia. *Parasit. Vectors* 8:610.

Rasic, G., et al. 2016. The *queenslandensis* and the type form of the dengue mosquito (*Aedes aegypti*) are genomically indistinguishable. *PLOS Negl. Trop. Dis.* 10:e0005096.

Raymond, M., et al. 1987. Detoxification esterases new to California, USA, in organophosphate-resistant *Culex quinquefasciatus* (Diptera: Culicidae). *J. Med. Entomol.* 24:24–27.

Raymond, M., et al. 1991. Worldwide migration of amplified insecticide resistance genes in mosquitoes. *Nature* 350:151–153.

Reddy, M. R., et al. 2007. Early evening questing and oviposition activity by the *Culex* (Diptera: Culicidae) vectors of West Nile virus in northeastern North America. *J. Med. Entomol.* 44:211–214.

Redman, S. N., et al. 2020. Linked-read sequencing identifies abundant micro-inversions and introgression in the arboviral vector *Aedes aegypti*. *BMC Biol.* 18:26.

Reidenbach, K. R., et al. 2009. Phylogenetic analysis and temporal diversification of mosquitoes (Diptera: Culicidae) based on nuclear genes and morphology. *BMC Evol. Biol.* 9:298.

Reidenbach, K. R., et al. 2012. Patterns of genomic differentiation between ecologically differentiated M and S forms on *Anopheles gambiae* in West and Central Africa. *Genome Biol. Evol.* 4:1202–1212.

Reidenbach, K. R., et al. 2014. Cuticular difference associated with aridity acclimation in African malaria vectors carrying alternative arrangments of inversion 2La. *Parasit. Vectors* 7:176.

Reisen, W. K., M. M. Milby, and R. P. Meyer. 1992. Population dynamics of adult *Culex* mosquitoes (Diptera: Culicidae) along the Kern River, Kern County, California, in 1990. *J. Med. Entomol.* 29:531–543.

Reisen, W. K., et al. 1991. Mark-release recapture studies with *Culex* mosquitoes (Diptera: Culicidae) in Southern California. *J. Med. Entomol.* 28:357–371.

Reiskind, M. H., and L. P. Lounibos. 2013. Spatial and temporal patterns of abundance of *Aedes aegypti* L. (*Stegomyia aegypti*) and *Aedes albopictus* (Skuse) [*Stegomyia albopictus* (Skuse)] in southern Florida. *Med. Vet. Entomol.* 27:421–429.

Reiskind, M. H., and M. A. Wund. 2009. Experimental assessment of the impacts of northern long-eared bats on ovipisiting *Culex* (Diptera: Culicidae). *J. Med. Entomol.* 46:1037–1044.

Reiter, P. 2007. Oviposition, dispersal, and survival in *Aedes aegypti*: implications for the efficacy of control strategies. *Vector-Borne Zoo. Dis.* 7:261–273.

Reiter, P., and D. Sprenger. 1987. The used tire trade: a mechanism for the world-wide dispersal of container breeding mosquitoes. *J. Am. Mosq. Contr. Assoc.* 13:494–501.

Reiter, P., et al. 1995. Dispersal of *Aedes aegypti* in an urban area after blood feeding as demonstrated by rubidium-marked eggs. *Am. J. Trop. Med. Hyg.* 52:177–179.

Reuben, R. 1973. The estimation of adult populations of *Aedes aegypti* at two localities in Delhi, India. *J. Comm. Dis.* 5:154–162.

Ribeiro, J. M. C. 1988. How mosquitoes find blood. *Misc. Publ. Entomol. Soc. Am.* 68:18–24.

Ribeiro, J. M. C. 2000. Blood-feeding in mosquitoes: proving time and salivary gland anti-haemostatic activities in representatives of three genera (*Aedes, Anopheles, Culex*). *Med. Vet. Entomol.* 14:142–148.

Ribeiro, J. M. C., and B. Arca. 2009. From sialomes to the sialoverse: an insight into salivary potion of blood-feeding insects. *Adv. in Insect Phys.* 37:59–118.

Ribeiro, J. M. C., and I. M. B. Francischetti. 2003. Role of arthropod saliva in blood feeing: sialome and post-sialome perspectives. *Ann. Rev. Entomol.* 48:73–88.

Ribeiro, J. M. C., and A. Spielman. 1986. The satyr effect: a model predicting parapatry and species extinction. *Am. Nat.* 128:513–528.

Richardson, J. B., et al. 2015 Evidence of polyandry in a natural population of *Aedes aegypti*. *Am. J. Trop. Med. Hyg.* 93:189–193.

Rigau-Pérez, J. G. 1998. The early use of break-bone fever (*Quebranta huesos*, 1771) and dengue (1801) in Spanish. *Am. J. Trop. Med. Hyg.* 59:272–274.

Ritchie, S. A., B. L. Montgomery, and A. A. Hoffmannn. 2013. Novel estimates of *Aedes aegypti* (Diptera: Culicidae) population size and adult survival based on *Wolbachia* releases. *J. Med. Entomol.* 50:624–631.

Rivera-Pérez, C., M. E. Clifton, and F. G. Noriega. 2017. How micronutrients influence the physiology of mosquitoes. *Curr. Opin. Insect Sci.* 23:112–117.

Rivet, Y., M. Marquine, and M. Ramond. 1993. French mosquito populations invaded by A2-B2 esterases causing insecticide resistance. *Biol. J. Linn. Soc.* 49:249–255.

Robich, R. M., and D. L. Denlinger. 2005. Diapause in the mosquito *Culex pipiens* evokes a metabolic switch from blood feeding to sugar gluttony. *Proc. Natl. Acad. Sci. USA* 102:15912–15917.

Rochlin, I., et al. 2019. West Nile virus mosquito vectors in North America. *J. Med. Entomol.* 56:1475–1490.

Rockefeller Foundation. 1938. *Annual Report* (New York: Rockefeller Foundation).

Romoser, W. S., et al. 1989. Histological parameters useful in the identification of multiple bloodmeals in mosquitoes. *Am. J. Trop. Med. Hyg.* 41:737–742.

Rosa, R., et al. 2014. Early warning of West Nile virus mosquito vector: climate and land use models successfully explain phenology and abundance of *Culex pipiens* mosquitoes in north-western Italy. *Parasit. Vectors* 7:269.

Rose, N. H., et al. 2020. Climate and urbanization drive mosquito preference for humans. *Curr. Biol.* 30:3570–3579.

Rose, N. H., et al. 2023. Dating the origin and spread of specialization on human hosts in *Aedes aegypti* mosquitoes. *eLife* 12:e83524.

Rosen, L. 1986. Dengue in Greece in 1927 and 1928 and the pathogenesis of dengue hemorrhagic fever: new data and different conclusions. *Am. J. Trop. Med. Hyg.* 35:642–653.

Ross, P. A., M. Turelli, and A. A. Hoffmann. 2019. Evolutionary ecology of *Wolbachia* releases for disease control. *Annu. Rev. Genet.* 53:93–116.

Rossignol, P. A., and A. Spielman. 1982. Fluid transport across the ducts of the salivary glands of a mosquito. *J. Insect Physiol.* 28:579–583.

Rossignol, P. A., J. M. C. Ribeiro, and A. Spielman. 1984. Increased intradermal probing time in sporozoite-infected mosquitoes. *Am. J. Trop. Med. Hyg.* 33:17–20.

Rossignol, P. A., J. M. C. Ribeiro, and A. Spielman. 1986. Increased biting rate and reduced fertility in sporozoite-infected mosquitoes. *Am. J. Trop. Med. Hyg.* 35:277–279.

Roux, O., and V. Robert. 2019. Larval predation in malaria vectors and its potential implicaiton in malaria transmission: an overlooked ecosystem service? *Parasit. Vectors* 12:217.

Rueda, L. M., et al. 1990. Temperature-dependent development and survival rates of *Culex quiquefasciatus* and *Aedes aegypti* (Diptera: Culicidae). *J. Med. Entomol.* 27:892–898.

Russell, B. M., B. H. Kay, and W. Shipton. 2001. Survival of *Aedes aegypti* (Diptera: Culicidae) eggs in surface and subteranean breeding sites during the northern Queesland dry season. *J. Med. Entomol.* 38:441–445.

Ryazansky, S. S., et al. 2024. The chromosome-scale genome assembly for the West Nile vector *Culex quinquefasciatus* uncovers patterns of genome evolution in mosquitoes. *BMC Biol.* 22:16.

Saarman, N., et al. 2017 Effective population sizes of a major vector of human diseases, *Aedes aegypti*. *Evol, Appl.* 10:1031–1039.

Saha, N., et al. 2012. Predation potential of odonates on mosquito larvae: implications for biological control. *Biol. Contr.* 63:1–8.

Samy, A. M., et al. 2016. Climate change influences on the global potential distribution of the msoquito *Culex quinquefasciatus*, vector of West Nile virus and lymphatic filariasis. *PLOS ONE* 11:e0163863.

Savero, O. P. 1955. Progress in the program for the eradication from the Americas of *Aedes aegypti*. Pp. 39–59 in *Yellow Fever: A Symposium in Commemoration of Carlos Juan Finlay*. Philadelphia: Jefferson Medical College of Philadelphia (Thomas Jefferson University).

Sawchuyk, L. A., and S. D. A. Burke. 1998. Gibraltar's 1804 yellow fever scourge: the search for scapegoats. *J. Hist. Med. Allied Sci.* 53:3–42.

Sayson, S. L., et al. 2015. Seasonal genetic changes of *Aedes aegypti* populations in selected sites of Cebu City, Philippines. *J. Med. Entomol.* 52:638–646.

Schmidt, H., et al. 2019. Transcontinental dispersal of *Anopheles gambiae* occurred from West African origin via serial founder events. *Commun. Biol.* 2:473.

Schmidt, T. L., et al. 2018. Fine-scale landscape genomics helps explain the slow spatial spread of *Wolbachia* through the *Aedes aegypti* population in Cairns, Australia. *Heredity* 120:386–395.

Schmidt, T. L., et al. 2020. Population genomics of two invasive mosquitoes (*Aedes aegypti* and *Aedes albopictus*) from the Indo-Pacific. *PLOS Negl. Dis.* 14:e008463.

Schoelitsz, B., et al. 2020. Chemical mediation of oviposition by *Anopheles* mosquitoes: a push-pull system driven by volatiles associated with larval stages. *J. Chem. Ecol.* 46:397–409.

Scolari, F., M. Casirghi, and M. Bonizzoni. 2019. *Aedes* spp. and their microbiota: a review. *Front. Microbiol.* 10:2036.

Scott, T. W., et al. 1993. Blood-feeding patterns of *Aedes aegypti* (Diptera: Culicidae) collected in a rural Thai village. *J. Med. Entomol.* 30:922–927.

Scott, T. W., et al. 2000. Logitudinal studies of *Aedes aegypti* (Diptera: Culicidae) in Thailand and Puerto Rico: blood feeding frequency. *J. Med. Entomol.* 37:89–101.

Scott, T. W., et al. 2006. DNA profiling of human blood in anophelines from lowland and highland sites in western Kenya. *Am. J. Trop. Med. Hyg.* 75:231–237.

Seenivasagan, T., et al. 2009. Electroantennogram, flight orientation, and oviposition responses of *Aedes aegypti* to the oviposition pheromone *n*-heneicosane. *Parasitol. Res.* 104:827–833.

Segev, O., R. Verster, and C. Weldon. 2017. Testing the link between perceived and actual risk of predation: mosquito ovposition site selection and egg predation by native and introduced fish. *J. Appl. Ecol.* 54:854–861.

Service, M. W. 1976. *Mosquito Ecology: Field Sampling Methods*. London: Applied Science.

Service, M. W. 1977. Mortalities of the immature stages of species B of the *Anopheles gambiae* complex in Kenya: comparison between rice fields and temporary pools, identification of predators, and effects of insecticidal spraying. *J. Med. Entomol.* 13:535–545.

Service, M. W. 1997. Mosquito (Diptera: Culicidae) dispersal—the long and short of it. *J. Med. Entomol.* 34:579–588.

Severson, D. W., et al. 1993. Linkage map for *Aedes aegypti* using restriction fragment length polymorphisms. *J. Hered.* 84:241–247.

Severson, D. W., et al. 1995. Restriction fragment length polymorphism mapping of quantitative trait loci for malaria parasite susceptibility in the mosquito *Aedes aegypti*. *Genetics* 139:1711–1717.

Severson, D. W., et al. 2002. Linkage map organization of expressed sequence tags and sequence tagged sites in the mosquito, *Aedes aegypti*. *Insect Mol. Biol.* 11:371–378.

Shah, S. 2010. *The Fever: How Malaria Has Ruled Humankind for 500,000 Years* (New York: Farrar, Straus & Giroux),

Shannon, R., and N. Davis. 1930. The flight of *Stegomyia aegypti* (L.). *Am. J. Trop. Med. Hyg.* 10:145–150.

Sheppard, P., et al. 1969. The dynamics of an adult population of *Aedes aegypti* in relation to dengue haemorrhagic fever in Bangkok. *J. Anim. Ecol.* 38:661–702.

Shocket, M. S., et al. 2024. Modelling the effects of climate and climate change on transmission of vector-borne disease. Pp. 255–320 in K. Fornace et al., eds., *Planetary Health Approaches to Understand and Control Vector-Borne Diseases*. Leiden, Netherlands: Brill Wageningen Academic.

Shousha, A. T. 1948. Species-eradication: the eradication of *Anopheles gambiae* from Upper Egypt 1942–1945. *Bull. World Health Organ.* 1:309–348.

Shroyer, D. A., and G. B. Craig. 1983. Egg diapause in *Aedes triseriatus* (Diptera: Culicidae): geographic variation in photoperiodic response and factors influencing diapause termination. *J. Med. Entomol.* 20:601–607.

Sih, A. 1986. Antipredator responses and the perception of danger by mosquito larvae. *Ecology* 67:434–441.

Silver, J. B., and M. W. Silver. 2008. *Mosquito Ecology: Field Sampling Methods*, 3rd edition. Dordrecht, Netherlands: Springer.

Simard, F., et al. 2009. Ecological niche partitioning between *Anopheles gambiae* molecular forms in Cameroon: the ecological side of speciation. *BMC Ecol.* 9:17.

Simoes, P. M. V., G. Gibson, and I. J. Russell. 2017. Pre-copula acoustic behaviour of males in the malarial mosquitoes *Anopheles coluzzii* and *Anopheles gambiae* s.s. does not contribute to reproductive isolation. *J. Exper. Biol.* 220:379–385.

Sinden, R. E. 2017. Targeting the parasite to suppress malaria transmission. *Adv. Parasitol.* 97:147–183.

Sinka, M. E., et al. 2010. The dominant *Anopheles* vectors of human malaria in Africa, Europe, and the Middle East: occurrence data, distribution maps and bionomic précis. *Parasit. Vectors* 3:117.

Sinka, M. E., et al. 2011. The dominant *Anopheles* vectors of human malaria in the Asia-Pacific region: occurrence data, distribution maps and bionomic précis. *Parasit. Vectors* 4:89.

Slotman, M. A., et al. 2005. Differential introgression of chromosomal regions between *Anopheles gambiae* and *An. arabiensis*. *Am. J. Trop. Med. Hyg.* 73:326–335.

Small, S. T., et al. 2023. Standing genetic variation and chromosome difference drove rapid ecotype formation in a major malaria mosquito. *Proc. Natl. Acad. Sci. USA* 122:e19835120.

Smithburn, K. C., et al. 1940. Aneurotropic virus isolated from the blood of a native of Uganda. *Am. J. Trop. Med. Hyg.* 20:471–473.

Soghigian, J., et al. 2020. Genetic evidence for the origin of *Aedes aegypti*, the yellow fever mosquito, in the southwestern Indian Ocean. *Mol. Ecol.* 29:3593–3606.

Soghigian, J., et al. 2023. Phylogenomics reveals the history of host use in mosquitoes. *Nat. Commun.* 14:6252.

Sol de Majo, M., P. Montini, and S. Fischer. 2016. Egg hatching and survival of immature stages of *Aedes aegypti* (Dipera: Culicidae) under natural temperature conditions during the cold season in Buenos Aires, Argentina. *J. Med. Entomol.* 54:106–113.

Soltis, P. S., R. A. Folk, and D. E. Soltis. 2019. Darwin review: angiosperm phylogeny and evolutionary radiations. *Proc. Biol. Sci.* 286:20190099.

Sommers, G., et al. 2011. Genetics and morphology of *Aedes aegypti* (Diptera: Culicidae) in septic tanks in Puerto Rico. *J. Med. Entomol.* 48:1095–1102.

Soper, F. L., and D. B. Wilson. 1943. Anopheles gambiae *in Brazil, 1930 to 1940*. New York: Rockefeller Foundation.

Sousa, C. A., et al. 2001. Dogs as a favored host choice of *Anopheles gambiae* sensu stricto (Diptera: Culicidae) of Säo Tomé, West Africa. *J. Med. Entomol.* 38:122–125.

Souza-Neto, J. A., J. R. Powell, and M. Bonizzoni. 2019. *Aedes aegypti* vector competence studies: a review. *Infect. Genet. Evol.* 67:191–209.

Southwood, T. R. E. 1978. Absolute population estimates using marking techniques. Pp. 70–129 in *Ecological Methods: With Particular Reference to the Study of Insect Populations*. Dordrecht, Netherlands: Springer.

Southwood, T. R. E., and P. A. Henderson. 2000. *Ecological Methods*, 3rd edition. Oxford: Blackwell Science.

Spielman, A., and J. Wong. 1973. Environmental control of ovarian diapause in *Culex pipiens*. *Ann. Entomol. Soc. Am.* 66:905–907.

Strand, M. R. 2018. Composition and functional roles of the gut microbiota in mosquitoes. *Curr. Opin. Insect Sci.* 28:59–65.

Subra, R. 1972. Études écologiques sur *Culex pipiens fatigans* Wiedemann, 1828 (Diptera: Culicidae) dans une zone urbaine de savane soudanienne ouest-africaine. *Entomol. Med. Parasitol.* 10:3–36.

Sumba, L. A., et al. 2008. Regulation of oviposition in *Anopheles gambiae* s.s.: role of inter- and intra-specific signals. *J. Chem. Ecol.* 34:1430–1436.

Swan, T., L. P. Lounibos, and N. Nishimura. 2018. Comparative oviposition site selection in containers by *Aedes aegypti* and *Aedes albopictus* (Diptera: Culicidae) from Florida. *J. Med. Entomol.* 55:795–800,

Tabachnick, W. J., and J. R. Powell. 1978. Genetic structure of East African populations of *Aedes aegypti*. *Nature* 272:535–537.

Tabachnick, W. J., and J. R. Powell. 1983. Genetic analysis of *Culex pipiens* populations in the Central Valley of California. *Ann. Entomol. Soc. Am.* 76:715–720.

Tabachnick, W. J., L. E. Munstermann, and J. R. Powell. 1979. Genetic distinctness of sympatric forms of *Aedes aegypti* in East Africa. *Evolution* 33:287–295.

Takken, W., and N. O. Verhulst. 2013. Host preferences of blood-feeding mosquitoes. *Ann. Rev. Entomol.* 58:433–453.

Tang, H., et al. 2005. Estimation of individual admixture: analytical and study design considerations. *Genet. Epidem.* 28:289–301.

Taylor, C. E., et al. 1993. Effective population size and persistence of *Anopheles arabiensis* during the dry season in West Africa. *Med. Vet. Entomol.* 7:351–357.

Tennessen, J. A., et al. 2021. A population genomic unveiling of a new cryptic mosquito taxon within the malaria-transmitting *Anopheles gambiae* complex. *Mol. Ecol.* 30:775–790.

Tewfik, H. L., and A. R. Barr. 1976. Paracentric inversion in *Culex pipiens*. *J. Med. Entomol.* 13:147–150.

Thawornwattana, Y., D. Dalquen, and Z. Yang. 2018. Coalesent analysis of phylogenomic data confidently resolves the species relationships in the *Anopheles gambiae* species complex. *Mol. Biol. Evol.* 35:2512–2527.

Thomas, C. J., D. E. Cross, and C. Bogh. 2013. Landscape movements of *Anopheles gambiae* malaria vector mosquitoes in rural Gambia. *PLOS ONE* 8:e68679.

Tirados, I., et al. 2006. Blood-feeding behaviour of the malarial mosquito *Anopheles arabiensis*: implications for vector control. *Med. Vet. Entomol.* 20:425–437.

Torr, S. J., et al. 2008. Towards a fuller understanding of mosquito behaviour: use of electrocuting grids to compare the odour-orientated responses of *Anopheles arabiensis* and *An. quadriannulatus* in the field. *Med. Vet. Entomol.* 22:93–108.

Touré, Y. T., et al. 1994. Ecological genetic studies in the chromosomal form Mopti of *Anopheles gambiae* s.str. in Mali, West Africa. *Genetica* 94:213–223.

Touré, Y. T., et al. 1998a. The distribution and inversion polymorphism of chromosomally recognized taxa of the *Anopheles gambiae* complex in Mali, West Africa. *Parassitologia* 40:477–511.

Touré, Y. T., et al. 1998b. Mark-release-recapture experiments with *Anopheles gambiae* s.l. in Banambani Village, Mali, to determine population size and structure. *Med. Vet. Entomol.* 12:74–83.

Tran Khanh, T., et al. 1999. *Aedes aegypti* in Ho Chi Minh City (Viet Nam): susceptibility to dengue 2 virus and genetic differentiation. *Trans. R. Soc. Trop. Med. Hyg.* 93:581–586.

Tripet, F., T. Thiemann, and G. C. Lanzaro. 2005. Effects of seminal fluids in mating between M and S forms of *Anopheles gambiae*. *J. Med. Entomol.* 42:596–603.

Tripet, F., Y. Touré, and G. Lanzaro. 2003. Frequency of multiple inseminations in field-collected *Anopheles gambiae* females revealed by DNA analysis of transferred sperm. *Am. J. Trop. Med. Hyg.* 68:1–5.

Trpis, M. 1972. Dry season survival of *Aedes aegypti* eggs in various beeding sites in Dar es Salaam area, Tanzania. *Bull. World Health Organ.* 47:433–437.

Trpis, M., and W. Hausermann. 1986. Dispersal and other population parameters of *Aedes aegypti* in an African village and their possible significance in epidemiology of vector-borne diseases. *Am. J. Trop. Med. Hyg.* 35:1263–1279.

Trpis, M., W. Hausermann, and G. B. Craig. 1995. Estimates of population size, dispersal, and longevity of domestic *Aedes aegypti aegypti* (Diptera: Culicidae) by mark-release-recapture in the village of Shauri Moyo in eastern Kenya. *J. Med. Entomol.* 32:27–33.

Tsuda, Y., et al. 2008. A mark-release-recapture study on dispersal and flight distance of *Culex pipiens pallens* in an urban area of Japan. *J. Am. Mosq. Contr. Assoc.* 24:339–343.

Tun-Lin, W., et al. 1996. Critical examination of *Aedes aegypti* indices: correlations with abundance. *Am. J. Trop. Med. Hyg.* 54:543–547.

Turelli, M., A. Katznelson, and P. S. Ginsberg. 2022. Why *Wolbachia*-induced cytoplasmic incompatibility is so common. *PNAS 119*:ewwqq637119.

Turner, T. L., M. Hahn, and S. V. Nuzhdin. 2005. Genomic islands of speciation in *Anopheles gambiae*. *PLOS Biol.* 3:e285.

Unger, M. F., et al. 2015. A standard cytogenetic map of *Culex quinquefasciatus* polytene chromosomes in application for fine-scale physical mapping. *Parasit. Vectors* 8:307.

Urbanelli, S., et al. 1981. Adattarnento all'ambiente urbano nella zanzara *Culex pipiens* (Diptera, Culicidae) [Adaptation to the urban environment in the mosquito *Culex pipiens* (Diptera, Culicidae)]. Pp. 305–316 in A. Moroni, O. Ravera, and A. Anelli, eds., *Ecologia, Atti XI Congressi nazionali*. Parma, Italy: Societa Italiana E. Zara.

Urbanelli, S., et al. 1997. California hybrid zone between *Culex pipiens pipiens* and *Cx. p. quinquefasciatus* revisited (Diptera: Culicidae). *J. Med. Entomol.* 34:116–127.

Urbanski, J. M., et al. 2010. The molecular physiology of increased egg desiccation resistance during diapause in the invasive mosquito, *Aedes albopictus*. *Proc. Biol. Sci.* 277:2683–2692.

Urbanski J. [M.], et al. 2012. Rapid adaptive evolution of photoperiodic response during invasion and range expansion across a climatic gradient. *Am. Nat.* 179: 490–500.

Urdoneta-Marquez, L., and A.-B. Failloux. 2011. Populaton genetic structure of *Aedes aegypti*, the principal vector of dengue viruses. *Infec. Genet. Evol.* 11:253–261.

van de Straat, B., et al. 2021. A global assessment of surveillance methods for dominant malaia vectors. *Sci. Rep.* 11:15337.

van Handel, E. 1993. Fuel metabolism of the mosquito (*Culex quinquefasciatus*) embryo. *J. Insect Physiol.* 39:831–833.

Vantaux, A., et al. 2021. Field evidence for manipulation of mosquito host selection by the human malaria parasite, *Plasmodium falciparum*. *Peer Community J.* 1:e13.

van Tol, S., and G. Dimopoulos. 2016. Influences of the mosquito microbiota on vector competence. *Adv. Insect Physiol.* 51:243–391.

Vavassori, L., et al. 2022. Multiple introductions and overwintering shape the progressive invasion of *Aedes albopictus* beyond the Alps. *Ecol. Evol.* 12:e9138.

Verna, T. N., and L. E. Munstermann. 2011. Morphological variants of *Aedes aegypti* collected from the Leeward Islands of Antigua. *J. Am. Mosq. Contr. Assoc.* 27:308–311.

Villoutreix, R., et al. 2021. Inversion breakpoints and the evolution of supergenes. *Mol. Ecol.* 30:2738–2755.

Vinayagam, S., et al. 2023. The microbiota, the malarial parasite, and the mosquito (MMM)—a three-sided relationship. *Mol. Biochem. Parasit.* 253:111543.

Vinogradova, E. B. 2000. Culex pipiens pipiens *Mosquitoes*. Sofia, Bulgaria: Pensoft.

Vonesh, J. R., and L. Blaustein. 2010. Predator-induced shifts in mosquito oviposition site selection: a meta-analysis and implications for vector control. *Israel J. Ecol. Evol.* 56:123–139.

Wallis, G. P., W. J. Tabachnick, and J. R. Powell. 1983. Macrogeographic genetic variation in a human commensal: *Aedes aegypti*, the yellow fever mosquito. *Genet. Res. Camb.* 41:241–258.

Wallis, G. P., W. J. Tabachnick, and J. R. Powell. 1984. Genetic heterogeneity among Caribbean populations of *Aedes aegypti*. *Am. J. Trop. Med. Hyg.* 33:492–498.

Wang, G., et al. 2021. Clock genes and environmental cues coordinate *Anopheles* pheromone synthesis, swarming, and mating. *Science* 371:411–415.

Wang, Y., et al. 2011. Dynamic gut microbiome across life history of the malaria mosquito *Anopheles gambiae* in Kenya. *PLOS ONE* 6:e24767.

Waterhouse, R. M., et al. 2018. BUSCO applications from quality assessments to gene prediction and phylogenomics. *Mol. Biol. Evol.* 35:543–548.

Watts, D. M., et al. 1973. Transovarial transmission of La Crosse virus (California encephalitis group) in the mosquito, *Aedes triseriatus*. *Science* 182:1140–1141.

Weaver, S. 2005. Journal policy on names of Aedine mosquito genera and subgenera. *Am. Soc. Trop. Med. Hyg.* 73:481.

Weitzel, T., et al. 2009. Genetic differentiation of populations within the *Culex pipiens* complex and phylogeny of related species. *J. Am. Mosq. Contr. Assoc.* 25:6–17.

Wekesa, J. W., R. S. Copeland, and R. W. Mwangi. 1992. Effect of *Plasmodium falciparum* on blood-feeding behaviour of naturally infected *Anopheles* mosquitoes in western Kenya. *Am. J. Trop. Med. Hyg.* 47:484–488.

Wesolowska, W., and R. R. Jackson. 2003. *Evarcha culicivora* sp. nov., a mosquito-eating jumping spider from East Africa (Araneae: Salticidae). *Ann. Zool.* 53:335–338.

White, B. J., F. H. Collins, and N. J. Besansky. 2011. Evolution of *Anopheles gambiae* in relation to humans and malaria. *Ann. Rev. Ecol. Evol. Syst.* 42:111–132.

White, B. J., et al. 2010. Genetic association of physically unlinked islands of genomic divergence in incipient species of *Anopheles gambiae*. *Mol. Ecol.* 19:925–939.

White, B. J., et al. 2016. Dose and developmental response of *Anopheles merus* larvae to salinity. *J. Exper. Biol.* 216:3433–3441.

White, G. B. 1985. *Anopheles bwambae* n. sp., a malaria vector in the Semliki Valley, Uganda, and its relationships with other sibling species of the *An. gambiae* complex (Diptera: Culicidae). *Syst. Entomol.* 10:501–522.

Wilke, A. B., et al. 2014. Population genetics of neotrophical *Culex pipiens*. *Parasit. Vectors* 7:468.

Wilkerson, R. C., Y.-M. Linton, and D. Strickman. 2021. *Mosquitoes of the World*, 2 vols. Baltimore: Johns Hopkins University Press.

Williams, C. R., et al. 2008. Rapid estimation of *Aedes aegypti* population size using simulation modeling, with a novel approach to calibration and field validation. *J. Med. Entomol.* 45:1173–1179.

Wint, W., et al. 2022. Past, present and future distribution of the yellow fever mosquito *Aedes aegypti*: the European paradox. *Sci. Total Environ.* 847:157566.

Woese, C. 1998. Default taxonomy: Ernst Mayr's view of the microbial world. *Proc. Natl. Acad. Sci. USA* 95:11043–11046.

Wondwosen, B., et al. 2017. A(maize)ing attraction: gravid *Anopheles arabiensis* are attracted and oviposit in response to maize pollen odours. *Malar. J.* 16:39.

Wong, J., et al. 2011. Oviposition site selection by the dengue vector *Aedes aegypti* and its implications for dengue control. *PLOS Negl. Trop. Dis.* 5: e1015.

Xia, S., M. L. Baskett, and J. R. Powell. 2019. Quantifying the efficacy of genetic shifting in control of mosquito-borne diseases. *Evol. Appl.* 12:1552–1568.

Xia, S., et al. 2020. Genetic structure of the mosquito *Aedes aegypti* in local forest and domestic habitats in Gabon and Kenya. *Parasit. Vectors* 13:417.

Xia, S., et al. 2021. Larval breeding sites of the mosquito *Aedes aegypti* in forest and domestic habitats in Africa and the potential association with oviposition evolution. *Ecol. Evol.* 11:16327–16343.

Xue, A. T., et al. 2020. Discovery of ongoing selective sweeps within *Anopheles* populations using deep learning. *Mol. Biol. Evol.* 38:1168–1183.

Yaro, A. A., et al. 2022. Diversity, composition, altitude, and seasonality of high-altitude windborne migrating mosquitoes in the Sahel: implications for disease transmission. *Front. Epidemiol.* 2:1001782.

Ye-Ebiyo, Y., R. J. Pollack, and A. Spielman. 2000. Enhanced development in nature of larval *Anopheles arabiensis* feeding on maize pollen. *Am. J. Trop. Med. Hyg.* 63:90–93.

Yen, J. H., and A. R. Barr. 1971. New hypothesis of the cause of cytoplasmic incompatibility in *Culex pipiens* L. *Nature* 232:657–658.

Yen, J. H., and A. R. Barr. 1973. The etiological agent of cytoplasmic incompatibility in *Culex pipiens*. *J. Invertebr. Pathol.* 22:242–250.

Yurchenko, A. A., et al. 2020. Genomic differentiation and intercontinental population structure of mosquito vectors *Culex pipiens pipiens* and *Culex pipiens molestus*. *Sci. Rep.* 10:7504.

Zahiri, N., and M. E. Rau. 1998. Oviposition attraction and repellency of *Aedes aegypti* (Diptera: Culicidae) to waters from conspecific larvae subjected to crowding, confinement, starvation or infection. *J. Med. Entomol.* 35:782–787.

Zamyatin, A., et al. 2021. Chromosome-level genome assemblies of the malaria vectors *Anopheles coluzzii* and *Anopheles arabiensis*. *GigaScience* 10:1–16.

Zapletal, J., et al. 2019. Predicting aquatic development and mortality rates of *Aedes aegypti*. *PLOS ONE* 14:e0217199.

Zembere, K., et al. 2022. The human-baited host decoy trap (HDT) is an efficient sampling device of exophagic *Anopheles arabiensis* within irrigated lands in southern Malawi. *Sci. Rep.* 12:3428.

Zhang, H., et al. 2022. A volatile from the skin microbiota of flavivirus-infected hosts promotes mosquito attractiveness. *Cell* 185:2510–2522.

Zhang, L., et al. 1996. An integrated genetic map of the African human malaria vector mosquito, *Anopheles gambiae*. *Genetics* 143:941–952.

Zhang, Q., and D. L. Denlinger. 2011. Molecular structure of the prothoracicotrophic hormone gene in the northern house mosquito, *Culex pipiens*, and its expression analysis in association with diapause and blood feeding. *Insect Mol. Biol.* 20:201–213.

Zhao, Z., et al. 2022. Mosquito brains encode unique features of human odour to drive host seeking. *Nature* 605:706–713.

Zouache, K., et al. 2022. Larval habitat determines the bacterial and fungal microbiota of the mosquito vector *Aedes aegypti*. *FEMS Microbiol. Ecol.* 98:1–11.

SUBJECT INDEX

Page references in *italics* indicate a figure; page references in **bold** indicate a table.

AUTHOR INDEX